The Economy of
British West Florida,
1763–1783

The Economy of British West Florida, 1763–1783

ROBIN F. A. FABEL

THE UNIVERSITY OF ALABAMA PRESS
Tuscaloosa and London

Publication of this book was made possible,
in part, by financial assistance from the
Andrew W. Mellon Foundation and the
American Council of Learned Societies, and by
a grant from the Auburn University Humanities Fund.

Library of Congress Cataloging-in-Publication Data

Fabel, Robin F. A., 1934–
The economy of British West Florida, 1763–1783.

Bibliography: p.
Includes index.
1. West Florida—Economic conditions. 2. West
Florida—Commerce—History. I. Title.
HC107.A13F32   1987   330.9759'902   86-4328
ISBN 0-8173-0312-X

British Library Cataloguing-in-Publication Data is available.

To Bob Rea

doyen of British West Florida historians

Contents

Acknowledgments

It is thanks to Robert R. Rea that I first became interested in British West Florida and that Auburn University has one of the better collections of books and microfilmed manuscripts on the province. Without his encouragement and wise advice, this book would not have been begun. Without the unfailing cooperation of the personnel of libraries, historical societies, and archives, it would not have been completed. Of prime help in successive summers was the busy staff of the Public Record Office, Kew. I received courteous assistance too in the Bodleian Library, Oxford; the Cabildo Archives, New Orleans; the Connecticut State Archives, Hartford; the Florida State University Archives, Tallahassee; the Georgia Historical Society, Savannah; the Georgia State Archives, Atlanta; the Library of Colonial Williamsburg, Williamsburg; the Linnaeus Library, London; the Mississippi State Archives, Jackson; the South Carolina Historical Society, Charleston; the Staffordshire County Records Office, Stafford; and the Tulane University Library, New Orleans. Above all I thank the helpful staff of the Ralph Brown Draughon Library at Auburn, especially in its microfilm center. I also gratefully acknowledge grants from Auburn University and from its Humanities Fund.

Part of chapter 3, in slightly different form, appeared in the proceedings of the Ninth Gulf Coast and Humanities Conference, and a condensed version of chapter 7 in the proceedings of the Tenth Wilburt S. Brown Conference on Military History. I am indebted to the editors of these publications for permission to use this material here.

For their benevolent interest in my work, I thank also several of my colleagues in the history department at Auburn University, including Gordon Bond, Wayne Flynt, Larry Gerber, Joseph Harrison, Hines Hall, and Larry Owsley, as well as the departmental secretaries, Flora Moss especially, who typed and retyped my manuscript.

The Economy of
British West Florida,
1763–1783

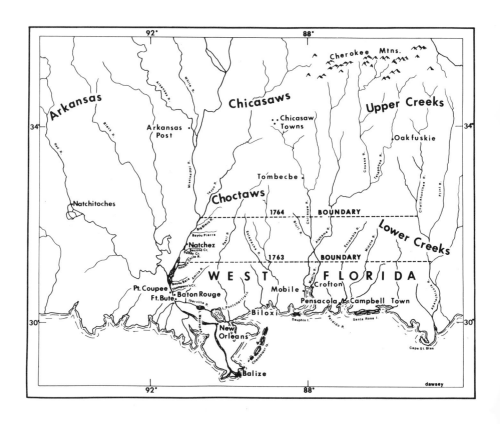

Introduction

Most Englishmen and even most Americans do not know that there was a British colony of West Florida in the 1760s and 1770s. That the province remained loyal in the revolution and, upon its conclusion, became not a state of the brave new American republic but instead part of the old Spanish empire helps to explain West Florida's absence from revolutionary mythology. When the British acquired it by diplomacy as a spoil of war in 1763, nobody could foresee that the colony's future would be brief. Many people believed that it would remain an unprofitable liability or, in contrast, would rival the older colonies farther north in population, prosperity, and political power.

During the Seven Years' War Florida had not been subjected to invading expeditions like those the British had launched against Havana, Manila, and Quebec. Little was known about Florida in Whitehall, although its vastness, which far exceeded that of the modern state of Florida, was so obvious that, very soon after British possession of Florida became sure, the British government decided to divide into two the new possession, which stretched all the way from the Mississippi to the Atlantic. Even so, each half was large, especially West Florida, which encompassed the panhandle of the modern state of Florida, about half of present-day Alabama, and a large proportion of what is now Mississippi in addition to some of Louisiana. This immensity was acquired in two stages. The original colony of West Florida, as defined in George III's proclamation of 7 October 1763, was

> bounded to the southward by the Gulf of Mexico, including all islands within six leagues [i.e., eighteen miles] of the coast from the river Apalachicola to Lake Pontchartrain; to the westward by the said lake, the lake Maurepas, and

the river Mississippi; to the northward by a line drawn due east from that part of the river Mississippi which lies in 31 degrees north latitude, to the river Apalachicola or Chatahouchee; and to the eastward by the said river.[1]

Less than a year later, the northern boundary of the province was extended considerably to embrace Natchez and the fertile area around it. The wording of the order in council which decreed the change seemed to indicate ignorance of the precise latitude of the new boundary, which was defined as "a line drawn from the mouth of the River Yasous [Yazoo] where it unites with the Mississippi due East to the River Apalachicola."[2] The result was to give West Florida a northern border at 32° 28'.

Some Britons thought the Floridas full of strategic and economic potential. Those who disagreed included opponents of the Bute ministry who found suspect anything achieved by the earl heading the administration which had negotiated the peace treaty by which the new colonies had been acquired. Others, perhaps without political affiliation, thought it strange that, having conquered developed Spanish possessions of known wealth, like the Philippines and Cuba but particularly Cuba, Britain had been persuaded to return them in exchange for an undeveloped region of unknown worth like Florida.

The Spanish did not think that the British had made a bad bargain. They were acutely conscious that their monopolistic domination of the Gulf of Mexico had been strategically shattered. In time of war the provinces of New Spain along its shoreline, and even Cuba itself, would be much more vulnerable to raids and invasion than they had been before Florida became British. In addition the Spanish commercial empire's vulnerability to penetration by British interlopers was enhanced by the cession of Florida, and this economic aspect most interested, not just British entrepreneurs, but government officials as well in the early 1760s, when war exhaustion distanced the prospect of renewed hostilities, although nobody imagined that Britain and Spain had fought their last war. That the Spanish had never made money from their Floridian possession was known but was of small concern. Eighteenth-century Englishmen had no respect for the commercial acumen of the Spanish nation.

The British had the self-confidence bred by a century and a half of establishing initially profitless colonies and then making them pay. And that undertaking—not uplifting the natives, nor cutting an imperial dash in bright uniforms, nor subjugating alien peoples—was the prime purpose of Britain's American empire in the 1760s. Lord Hillsborough became secretary of state for the American colonies in 1768. The title of his office, often ambiguously shortened to "American secretary," will in this work be given as "plantations secretary," which was an unambiguous contemporary usage. He defined the principal raison d'être of the continental colonies as "giving proper encouragement to the fishery, to the production of naval stores, and to the supply of the sugar islands with lumber and provisions. . . . I cannot conceive that it can ever be sound policy in this kingdom to allow settlement in places where none of the above named great natural objects are attainable."[3]

Hillsborough's was a limited vision, but West Florida conformed quite well to his criteria. It never did much for the great northern fishing grounds, which is what Hillsborough no doubt had in mind, but it was a maritime province supporting numbers of vessels which did gather fish, mollusks, and turtles for local consumption from the Gulf of Mexico, one of the most abundant fisheries in the world. The pine forests of the province also could and would provide ships' stores, tar, turpentine, masts, and spars. West Florida also had the potential for supplying the British West Indies with lumber and did so in the later 1770s. The trade might well have endured had Britain been able to retain the Floridas after 1783.

Not everybody believed that West Florida had such potential. "Junius Americanus," in a critical open letter to the earl of Hillsborough in 1769, wrote that the Floridas were "in America as Arabia in Asia" and likely to prove useful to criminals on the run but never "useful or advantageous to the state."[4]

Equally critical was the "Pennsylvania Farmer" John Dickinson. His argument was tendentious. He was less concerned with the alleged worthlessness of the Floridas than with protesting British taxation. Nevertheless he was probably reflecting and playing upon an accepted opinion of Florida when he wrote that "the British colonies are to be drained of the rewards of their labour to cherish the scorching lands of Florida . . . , which will never return to us one farthing that we send them."[5]

To govern any American colony in the 1760s and 1770s successfully was at the very least extremely difficult. From its creation in 1763 to its de facto extinction as a British possession in 1781, West Florida had half a dozen chief executives. Only three of them were formally appointed; the others acted as governors without the title. All of them—Robert Farmar, George Johnstone, Montfort Browne, John Elliott, Elias Durnford, and Peter Chester—were officers in King George's armed forces, revealing how the ministry correctly viewed the new colony as a pioneer settlement whose stability was threatened in various ways.

The warrior Indian tribes within the borders of West Florida were likely to resent encroachment on their hunting grounds. They heavily outnumbered the white settlers, who included the misfits and rogues usually attracted to

new frontier settlements; Governor Johnstone repeatedly called them the
overflowing scum of empire. Their existence alone might have suggested
that it would be prudent to appoint a military man to the governorship, but,
in addition, another segment of the population was French. It was reasonably
supposed, although without result, as it turned out, that they might be
restless and uncertain in their loyalty to the union jack. Externally the main
threat was the Spanish of Louisiana. Once, after considerable delay, they had
finally occupied the province, their forts and comparatively numerous garrisons were separated from West Florida merely by the width of the Mississippi River. All these elements, both internal and external, were genuine
potential menaces, but another chronic problem for the governors was political malcontents. It would have been extraordinary indeed if men immi-
grating from New England, New York, Pennsylvania, and South Carolina
had not brought with them to West Florida some of the revolutionary ideas
fermenting in their native colonies in the 1760s and 1770s. They were balanced, it is true, by individuals who were apolitical or calculatedly neutral or
outspokenly loyal. But from the tenure in office of George Johnstone, West
Florida's first civil governor, who had a veritable gift for creating division,
right up to Peter Chester, the last, who, like King Charles I, summoned
assemblies of the people's representatives only when he had no alternative,
there was disunity.

The basis of political rivalries in West Florida is not simple to characterize.
Personal attachments and desire for office, as well as a popular clamor for
curbing the authoritarianism of the king's representatives, all played a part,
but self-interest was probably more important, both for those upholding
royal authority and for those contesting it, than any abstract principle. However the dissidence may be explained, the result was turbulence. Governor
James Grant of East Florida commented on "that spirit of dissension which
rages all over America nowhere more than in West Florida where they are
. . . as full of faction, dispute and politics as they are at Boston or
Charleston."[6]

No doubt a good number of the men and women who lived in West
Florida during the British period had political opinions and prejudices, but
they did not migrate there for political reasons until the tail end of the period,
when there was some, limited, loyalist immigration. Nor, except as soldiers,
sailors, or administrators, were they in the province for any public purpose.
They were motivated by the impulse that lay behind much of American
immigration history—the hope for, belief in, and prospect of economic
betterment.

The vision of improvement took various forms. Some hoped to make a fortune selling to Spanish neighbors or to buy at huge profit from Indians; others more modestly sought their living supplying the needs of West Florida's garrison or those of incoming colonists. Some hoped to establish plantations on which they could grow on a large scale crops which they had grown on a smaller scale elsewhere to meet local or imperial needs; others thought that in West Florida it might be possible to produce more exotic items difficult or impossible to obtain elsewhere from Britain's possessions, such as wine, silk, and quassia. Some entrepreneurs purveyed services, making available their ships or their skills as masons, bakers, or carpenters. A surprising number of low-caliber lawyers, having failed elsewhere, sought to practice their profession in West Florida. A considerable contingent hoped to profit from dealings in real estate. An even greater number made a living by combining different types of profitable activity.

The object of the present book is to consider how these people achieved or failed to achieve their ambitions, whether the province as a whole was intrinsically economically viable, and whether the generally held belief that West Florida was an economic failure is a fair judgment. A secondary aim is to analyze some aspects of the subject which have been comparatively neglected in the two older standard works on the economic life of British West Florida.[7] Examples would include the maritime life of the province, the institution of slavery in West Florida, and the potentially great immigration scheme sponsored by the Company of Military Adventurers, which last would no doubt have been treated by Clinton Howard had he continued his study of the economic development of the province beyond 1769.

I

Immigration

There were two distinct waves of immigration to West Florida. The first
followed a flurry of publicity, including offers of land, in the mid-1760s and
attracted a considerable number of settlers from Europe as well as from
elsewhere in America. There followed a lull in the late 1760s and early 1770s,
when, perhaps not coincidentally, the number of vessels visiting West Florida
was half that frequenting its ports in 1765 (see Appendix 1). Interest in living
in West Florida picked up again in 1772 and 1773 as news spread of the
superior land to be had on or near the Mississippi, and a second wave of
immigration, mostly of Americans, followed, continuing through the early
years of the revolution.

Everybody connected with the province except, naturally, the indigenous
tribe of Indians saw the desirability of attracting permanent settlers. It was
clearly in the strategic interest of the crown to persuade colonists to migrate
there from the homeland or from other British possessions, because West
Florida was a frontier province, adjoining the potentially hostile Spanish
empire and within striking distance of any conquering expedition which
might be launched from Spanish Cuba. Its extension westward also facilitated
British influence over the Indians of the interior, which could be an important
advantage in time of war. The government in London let it be known in
British and American newspapers that an abundance of land was either freely
or cheaply available in West Florida. British nationality was not a mandatory
qualification.[1] French, German, and Irish settlers emigrated to West Florida,
but the main appeal was to Britons born on both sides of the Atlantic.

"The great object in every colony," wrote the author of "Thoughts concerning Florida" to the secretary responsible for the colonies, was "the encouragement of population." The "Thoughts" is anonymous but was almost
certainly written at the instigation, if not by the hand, of George Johnstone.[2]
It argued that existing inhabitants of the Floridas, left over from the years of

Bourbon rule, should be dissuaded from leaving if their reason for doing so was simply that they would have to live under the British flag. The paper argued too that the reduction of the armed forces at the end of the Seven Years' War offered an unusual opportunity to recruit a number of good settlers for West Florida. The author suggested that 800 might be a suitable number of males and that they should be complemented by 500 females from Britain. Overseers who understood plantation life might be attracted from Jamaica. It was hinted as well that, if Johnstone were appointed governor of West Florida—and his appointment was probably the main purpose of the document—he could personally recruit 300 settlers.[3] Johnstone subsequently obtained his governorship but in fact enlisted only a handful of emigrants.

Nevertheless the principle elaborated in the "Thoughts" was approved by the government. Johnstone's gubernatorial instructions stressed the benefits of settling his province swiftly. They touched on the desirability of enhancing British government revenues through such settlement, but no oppressive intent need be suspected. The infant colony was being launched on parliamentary appropriations; to hope for a time when the government subsidy could be reduced and eventually eliminated was only sane.

Because in time past governors of other colonies had awarded large tracts which were never subsequently cultivated, Johnstone was ordered to discriminate in favor of applicants for land who seemed best able to improve their grants by settling on them as many white servants and blacks as the acreage required. Taking a stance that contrasted with ministerial policy when Georgia had been founded thirty years before, the government countenanced slavery in West Florida.

The governor was told the three-step procedure for granting land. He would first issue a warrant signifying approval of a petition of a grant in a particular area to the provincial surveyor general, the approval having come, not at his unsupported whim, but as a result of deliberation with his council. As the warrant was issued, dockets would be placed in the offices of the registrar and auditor of West Florida to indicate to other land seekers that the land in question was provisionally allocated. Either in person or through subordinates, the surveyor general would survey the land and within six months would return the warrant together with a plat or description of the surveyed land. The governor in council would then formally approve the land grant by issuing a patent registering it, all conditions of tenure having been written into the patent in the registrar's office, with simultaneous notification to the auditor's office.

Any head of a family could apply for 100 acres for himself or herself. In addition he or she was entitled to 50 acres for every member of his or her "family," a word which included not merely spouse and offspring but also white indentured servants and black slaves. If the governor approved, an applicant could supplement this free "family right" land by buying up to 1,000 additional acres at the rate of five shillings for 50 acres.

The conditions on which land was granted were detailed and stringent. For every fifty acres of plantable land granted, at least three acres had to be cleared and worked within three years. If the land was barren, three cattle had to be placed on it for every fifty acres. Holders of uncultivable land could also prove their serious intent by employing at least one servant to dig a quarry or mine and by building a dwelling measuring at least twenty feet by sixteen.

Once a grantee had improved the land in accordance with these conditions, he or she was advised to show proof of improvement at the nearest local court of law and to have the proof certified in the provincial register's office in order to escape the possible punishment for failure to observe the conditions upon which the land had been received, which was forfeiture. In practice this penalty was largely theoretical, although a few cases of forfeiture can be found in the governorship of the energetic Johnstone.

To make it possible to fulfill the conditions, Johnstone was instructed to take care that the surveyors specified, when making plats, which land was plantable and which uncultivable. He was also told to be equitable in his distribution of good and bad land and, where river land was concerned, to ensure that the land grants should have a breadth not exceeding one-third of the length and that the breadth, not the length, should lie along the valued river shore. A final condition of receiving a land grant was agreement to pay a quitrent of a halfpenny for every acre. No quitrent was due for at least two years after a land grant had been made, but on the first Michaelmas day following the two-year period, and on every Michaelmas day thereafter— Michaelmas day being the feast of St. Michael, 29 September, one of the four quarter days of the British business year—the quitrent was legally payable. In fact there is no record that anyone ever paid a quitrent in West Florida.[4]

In spite of the minutely detailed regulations with which these family right grants were defined, the government's intentions were liberal. The government was generous too in the grants of free land which could be claimed by veterans of the Seven Years' War, who, under the terms of the royal proclamation of October 1763, could qualify for tracts of various size, according to military rank, details of which will be given elsewhere in this work. In the case of proclamation grants, frivolous applications were discouraged by a requirement that noncommissioned veterans had to live in America and had to apply in person in order to receive land.[5] This guarded liberality was justifiable; to strengthen the empire at its edges by encouraging veterans to settle there served the national interest of Great Britain.

Less well advised was the alacrity with which, after the foundation of West Florida, large mandamus grants were initially made on royal orders to certain gentlemen who enjoyed the favor of the king, even if they were not resident in North America, and the obvious purpose of these grants was profit from land speculation rather than the organization of emigration schemes. Such largesse was of course at variance with the caution that Johnstone was urged to show in disposing of sizable tracts. Under this contradictory policy such nonresidents as Lord Elibank, Lord Eglinton, and Samuel Hannay all received grants of thousands of acres.

To explain this aberration, comment on the contemporary background of British domestic history is necessary; no digression is involved, since the story of British colonization of the eighteenth century was so heavily affected by political influences, particularly in the 1760s. A cause célèbre of the day was the alleged sway exerted over the sovereign by his former tutor, the Scottish earl of Bute, whom the young George III had made first lord of the treasury in May of 1762, to the consternation and outrage of experienced politicians. The result, ran their complaint, was a sinister scotchification of government and Bute's continued influence over the king even after he surrendered the seals of political office in April 1763.

West Florida provides some evidence for these charges. George Johnstone was a Scot and a close friend of John Home, Bute's secretary. He was appointed governor seven months after Bute's nominal fall. One of Johnstone's uncles, James Murray, was appointed governor of Canada at the same time. Another uncle, Lord Elibank, was given a mandamus grant of 20,000 acres near Pensacola. Samuel Hannay, a Scottish businessman and a crony of Johnstone's, received a mandamus grant of 5,000 acres on the Mississippi.[6] He became provost marshal of West Florida but, since he preferred life in London, found it convenient to delegate his duties to his relation John Hannay. The earl of Eglinton was a dining companion of Johnstone's and yet another Scot.[7] He was given a mandamus grant of 20,000 acres near Pensacola.[8] Another intimate of Johnstone's who never went near West Florida was his future political ally George Dempster, who obtained a 20,000-acre grant.[9] These mandamus grants were made years after Bute had ceased to be prime minister. Their owners made no serious attempt to settle them, but the mandamus grants made to Montfort Browne and Phineas Lyman, both of whom were residents of North America and who made strenuous efforts to people them, fell into a different category. Each received 20,000 acres somewhat later than the Scots mentioned above in the rich, important, and, as the events of 1779 would show, strategically vulnerable western portion of the province.[10]

Strategy apart, West Florida's need for settlers had economic and social dimensions. The merchants of Pensacola and Mobile were glad to see their customers increase. The garrison troops and their officers no doubt viewed immigrants as improving the bare social life of the remote colony. The crown officials, especially the governors, saw that the colony's prosperity and perhaps even its survival depended on the attraction of settlers. It was not merely a question of general prosperity. Common practice and good taste dictated that officials should write in terms of "the interests and safety of His Majesty's dominions," but personal prosperity also depended on settlement. Almost all in the colony, whatever their social standing, including governors, customs officers, soldiers and sailors, merchants, bakers, and carpenters, had bought land or had received it free. An increase in population meant capital appreciation.

Having arrived in Florida late in 1764, Governor Johnstone described in glowing phrases the attractions of his province in the British and colonial press.[11] Perhaps because he was a Scot, but more likely because of observations he had made in the West Indies when he was a naval officer, Johnstone did not rate Englishmen highly as immigrants, preferring Frenchmen and Germans.[12]

No doubt he had spoken enthusiastically about West Florida even before he sailed aboard the storeship *Grampus* from Portsmouth in June 1764, for he was accompanied not only by bags of seeds of all sorts, household furniture for the settlers, and military stores but also by several workmen in the building trade.[13] Although the numbers were always fewer than expected, many settlers arrived in West Florida during Johnstone's governorship. They were of various nationalities, but the bulk came from Britain and her possessions.

The London government supported them in various ways. Late in April 1764 it was reported that a government-chartered sloop carrying fifty emigrants for the province had put into Charleston on her way to West Florida.[14] The government also encouraged free-lance emigration to West Florida. Unfortunately, the attempts of the board of trade, governors, and lieutenant governors to portray the province as a pleasant and profitable place were balanced by press reports both factual and fanciful which painted a grimmer picture. Other reports reminded readers of the peril of sea voyages.

An early blow to settlement had been the heavy loss of life among prospective immigrants when the *Robert and Betty* from Liverpool foundered in October 1763. She had called at Madeira to take on wine and water when a nocturnal storm had driven her from her anchors onto rocks in the harbor. Of the 213 passengers, appallingly, 195 perished. Since the passengers were all Protestants, the victims were denied the right of interment in Madeiran soil and thus were buried at sea. It was a grievously unfortunate start to Floridian immigration. Disillusion can also be read in the report that a brigantine arrived in Savannah in June 1765 bringing twenty-one settlers *from* West Florida.[15]

Another discouraging report came from the pen of an anonymous writer who came to Pensacola on 1 September 1764, one of over a hundred immigrants who had been led to believe that West Florida was a "terrestrial paradise." He wrote, "Alas, how were all their hopes blasted when they arrived and instead of the finest country in the world, found the most sandy, barren and desert land that eyes could see or imagination paint." This judgment, part of a much longer letter of condemnation, appeared in the *Scots Magazine* of January 1765. Its effect on would-be emigrants in the British Isles may easily be guessed. From neighboring Mobile another newcomer in the following year wrote that West Florida was "good for nothing but destroying Englishmen."[16] A different kind of bad publicity for the Floridas, doubtless effective to some extent in discouraging would-be emigrants, was the repeatedly found tale that prostitutes and female convicts were being exported there. One report stated, "Since Monday last 43 nymphs of Covent Garden have been engaged to visit Pensacola in West Florida on high encouragement where, it seems, there is at present a great want of the sex."[17] That there was a serious intention to send females from the sump of British society may have been more than a good newspaper story. In 1766 there was a mutiny aboard a convict ship from Dublin. The mutineers killed the captain and crew and fled ashore, abandoning the transport and a number of female convicts who were helpless with fever. They had allegedly been destined for East Florida.[18] This horrific fiasco perhaps dissuaded the government from persisting in a policy of transporting criminals to the Floridas.

Not all attempts to bring in Europeans were abortive, although the numbers of those who actually immigrated, as reported in the press, were probably inflated. From London came the news on 5 August 1764 that a great number of stonemasons had left Liverpool to build wharves at Pensacola.[19] On 19 April 1765 it was alleged that thirty bakers had been engaged to work in the Floridas.[20] On 25 August a man who had been granted a large tract of land in Florida was said to have hired two Germans to recruit settlers in their native land. Even at that very time, according to report, an English ship was in the Elbe embarking passengers from the Palatinate for West Florida.[21] Two years later an additional German emigration to West Florida was reported, this time specifically of peasants skilled in viticulture from the Rhineland. Despite the repetition of such accounts, there are few German names to be found among the permanent residents of the province, which is surprising, since we know in detail of one sizable influx of German immigrants. They were a cargo of forty-five servants imported by and subsidized by Jeremiah Terry and Company. They came in the ship *Franklin,* Isaac Lassalls captain, from Rotterdam to Charleston. The next leg of the voyage, to Pensacola, was in the sloop *London,* commanded by Jonathan Clark.[22] These immigrants arrived in May 1766, which no doubt explains why Jeremiah Terry, in August, received a large grant of 4,050 acres in West Florida.[23] In theory the adults bound themselves to serve Terry for four years, and the children to serve until they were twenty-one. The scheme remained theoretical. Either disease decimated the immigrants or, more likely, they continued emigrating to somewhere else, because their names do not recur in the documents relating the history of West Florida.

Similarly disappointing in their outcome were the schemes of Johnstone's lieutenant governor, Montfort Browne, to bring into West Florida numbers of French and Irish settlers. Gifted, it seems, with some useful connections and undoubted powers of persuasion, Browne received a generous 20,000-acre grant that was to be located in any part of the province of West Florida that he chose. From England he sent his brother William to Pensacola to obtain the land while he himself stayed in the British Isles to organize the labor force to work it. William did not manage his responsibility well. He obtained a grant from the West Florida council of Dauphin Island, a very unsatisfactory choice for at least three reasons: the whole island contained only 2,600 acres; its possession was contested by an older inhabitant; and it was not suitable for the location of several hundred feudal minions (which Browne seems to have had in mind). Not until December 1767, when Browne was in the influential position of acting governor, could he successfully assert his claim to the remaining 17,400 acres of the original royal grant.[24] Prior to that development, in 1765, Montfort Browne was recruiting immigrants.

Full details of Browne's emigration—some called it kidnapping—scheme have never come to light. Part of the complement of passengers on the emigrant ship which Browne chartered were Huguenots. Their forlorn attempt to settle in Campbell Town near Mobile has been reasonably well documented because their emigration to West Florida was heavily subsidized. The British government supplied them with free ship passage, clothing, tools, firearms, and a subsistence allowance for at least nine months, presumably until the first harvest was in. No doubt the contract to supply the Huguenots with provisions was much sought after. It went to a Robert Ross who later left West Florida for New Orleans.[25] Lieutenant Governor Montfort Browne organized them and accompanied them to Florida with their leader, the Reverend Peter Levrier, who received an annual government stipend of £100. In addition the board of trade instructed Governor Johnstone to let the French Protestants choose 20,000 acres of crown land in his province. There were high hopes that the Huguenots, who had a reputation as thrifty hard workers, would be a major asset to West Florida's economy. They proved to have no understanding of the cultivation of silk and wines and were but indifferent planters. As early as June 1768 it was reported to the plantations secretary that they had been reduced to a very small number.[26] By 1770 Campbell Town had become a "rotten borough," with representatives in the legislature but no voters. The sorry misadventures of the Huguenots on the banks of the Escambia have been well described by J. Barton Starr.[27] Initially they numbered only forty-eight, but with them to Florida came a more numerous but less noticed body of Irish men and women. No existing history of West Florida mentions them.

The Irish contingent was part of a private scheme devised and paid for by Montfort Browne for his own profit. Having accepted no government money for them, he was under no obligation to return accounts to the board of trade, and he did not even mention them in his official correspondence. There are two scanty references to them in contemporary newspapers. One noted in 1766 that Browne had arrived at Pensacola with 200 settlers from Ireland.[28] In the following year another recorded that a subscription had been opened at Pensacola for the relief of sick Irish settlers and that £200 was raised almost at once, the funds being administered by the parish churchwardens.[29] We know more only because a compassionate South Carolinian visited Mobile in July 1766 and recorded his indignation at the plight of the "Hibernians" he found there. He asked the reader to picture "a number of men, women and children laying [*sic*] and strolling about the streets, some dead, some dying, some unsuccessfully begging, and others shaking in an ague, or burning in a raging fever."[30] Upon inquiry the author found that these unfortunates all believed that they had been swindled ("trepanned" was the contemporary slang word actually used) by Montfort Browne, who had gulled them with a fine-sounding scheme published under the title *Advantageous Proposals*. Alleging that he had recently returned from America, which was true, and was well acquainted with West Florida, which was not, Browne, in the *Proposals,* had painted a golden picture of the province. He had assured readers that industrious immigrants were guaranteed not merely a very comfortable livelihood but even a fortune. To any person prepared to sign a three-year contract to work on his land, Browne promised free passage, 100 acres, and enough seed corn to sow them. To those who paid their own passage, Browne offered as many acres as they wished. If any of his indentured servants fell ill in Florida, Browne promised not to hold them to the agreed years of service.

The emigrants sailed on the *Red Head* galley from Cork. Once aboard and at sea, all those who had not signed indentures and other types of legal servitude were pressed into doing so. Evidently Browne, if he could not use them himself, wanted them in bondage so that he could sell their services to West Floridians who needed help in their businesses or on their plantations. On arrival in Pensacola no one sought such servants, either because slave labor was cheaper or because of poverty or disgust with Browne. As a result Browne absolved himself from further responsibility and allowed his charges to fend for themselves. The anonymous publicizer of their plight, who also reproduced the *Advantageous Proposals,* roundly denounced Browne as a "kidnapper" and "trafficker in human flesh." The fate of his victims is unsure, although "a great many" died after their first experience of West Florida's sickly summer. One may hazard a guess that, by comparison, Levrier's Huguenots were lucky, for they at least had a daily subsistence allowance of fourpence from crown funds.[31]

In spite of such disappointments and unfavorable publicity, in April 1768 a traveler to New England from West Florida reported that Pensacola was settling fast from the daily arrival of Europeans, so that over an eighteen-month period nearly 200 houses had been built there.[32] More promising, as it turned out, than the influx of Europeans, who often fell prey to the diseases of a semitropical climate to which they had no resistance, was immigration from other American colonies. As soon as it was known, in the early 1760s, that French Louisiana was to be placed under the rule of the king of Spain, there were British predictions that disgruntled French habitants would prefer to live in neighboring British West Florida, which sometimes came true. For example Governor Johnstone persuaded François Caminada, a distinguished citizen of New Orleans, to move to Pensacola, where he became a member of the provincial council in 1765,[33] and Lieutenant Governor Montfort Browne wrote with satisfaction of having administered allegiance oaths to "a good many" Frenchmen disgusted with Spanish rule. The French migration, however, was a trickle rather than a flow.[34]

In the spring of 1768 it was reported that a Virginian named James had arrived in West Florida with five others who owned considerable land in his native state. They decided to apply for a large tract on the Escambia near Pensacola.[35] As so often happened, announced intention was not followed by action. Many reasons can explain such failure, ranging from sickness and the

steep cost of legal and surveying services to incidents involving hostile Indians. Some such incidents actually occurred, although rumors on the subject
were also rife and tended to deter potential settlers. William Bartram cited
in his travel book the Lower Creek murder of white families at Apalache Bay
who were migrating from Georgia to Mobile.[36]

In general that part of the province that lay along the Mississippi, Amite,
and Comite rivers proved more attractive to planters than more easterly
regions, the poverty of whose soil had early been discovered and publicized.[37]
Once the superiority of western land was realized, there was a rush of settlers
such as Pensacola and Mobile had never seen. "Inhabitants from all quarters
are continually flocking to it," commented the *Providence Gazette* of 24 July
1773. In February 1774 a Pensacolan wrote that "the Mississippi goes on
with so rapid a progress that it astonishes everyone who does not know the
country." He expected within a few weeks the arrival of thousands of settlers
from the Company of Military Adventurers, an emigration organization so
significant that I treat it below in a separate chapter.[38] But even before the
Adventurers began to arrive, other schemes brought in substantial numbers
of immigrants. Captain Amos Ogden of New Jersey was given permission
to reserve a strip of 15,000 acres near Natchez for immigrants in October
1772, and he brought in enough by September 1773 to qualify for 3,550
acres.[39] From the same colony Samuel Sweezy arrived with sixty-six settlers
in the spring of 1773.[40]

On the eve of the revolution, impetus was given to immigration by the
publication of Bernard Romans's history of the Floridas, of which an unde-
clared purpose was to attract settlers. His detailed advice to a prospective
immigrant is useful in revealing conditions in West Florida in the 1770s. The
information is overoptimistically presented, but at least we may learn the
approximate expenses facing an immigrant and gain some idea of prevailing
prices. From a northern or middle colony, a vessel might be chartered at the
rate of $2½ a day. Supposing that the passage to the Mississippi region of
West Florida took sixty days and that the vessel suitable for a man, his family,
and fourteen slaves was sixty tons' burden, then the charter fare alone would
cost $300. Provisions consumed on the voyage by this group, on the as-
sumption that the family included four children, together with the cost of
buying and feeding an initial stock of pigs, sheep, and poultry, would cost
another $40. Romans estimated that wine and rum supplies would cost the
surprisingly high figure of $150, while axes, saws, spades, and hoes would
cost another $60. He also assumed that this hypothetical emigrant would
acquire 1,850 acres by a combination of headright and purchase. Of these
acres 850 would be free headright acres, obtainable because he, his wife, his

four children, and ten of his slaves would be permanent settlers (four slaves would be sold); the remaining 1,000 acres would have to be bought.

Although the land that was not free would be extremely cheap—ten shillings, or just over $2, per hundred acres—the necessary fees payable were considerable. Apart from the comparatively minor but numerous fees for petitioning, for the warrants for the precepts and the plat, a substantial fee went to the surveyor general, who charged $1 for every hundred acres surveyed. More substantial still was the $20 payable to the governor for his signature and the provincial secretary for his book work. In addition, the potential planter would have to pay to have the land physically surveyed. All expenses and fees arising from such a sizable tract would thus amount to $155. Erecting a house for himself and his family for four months, after which, estimated Romans, the first crops could be harvested, would run to another $150. Stocking the farm with six cows, six hogs, some poultry, and a horse would cost $136, while buying an indigo vat and a boat would take $55.

Thus, without taking into account delays and accidents, the would-be planter would have to invest $1,296, a very considerable sum for the day. Such an expenditure would exclude all but the well heeled, but Romans suggested that a man intending to emigrate with a wife and four children and a slave could contemplate becoming a planter even with as little as $400 capital if he would be content with 400 acres, make his own canoe, and take passage on, rather than charter, a vessel. Emigrants with less than $400 should be content with selling their labor.

Those with capital resources were doubly blessed because they could take advantage of the disparity between the price of slaves in the northern colonies and in the new one. Romans advised the prospective planter to import more slaves than he needed to the new province, for he could sell the surplus at a handsome profit. Young boys and girls could be bought in the north for $100 each. They could each be sold on arrival at the Mississippi for $150.

We further note that Romans's suggestions contain a good deal of incidental information about life in the 1770s. His tone is optimistic, but he wanted to imply that success in West Florida came only through hard work. Ostensibly he discouraged the dandy or the socialite: "I write not for him." At the same time, he suggested that survival for the white drifter was not too difficult in the colony. He advised emigrants to rely on black slaves for labor because any white servants brought for that purpose would within three months resent having to call anyone master. Such a servant, he reasoned, would be corrupted by the abundance of white idlers already in the province. The real reason probably lay in the economic environment. There was a superfluity of free land. Blacks could not qualify for it, but any white servant ambitious to achieve self-sufficiency would naturally resent working out the five or seven years specified by indentured servitude on another man's land, knowing that, except for his indentures, he too could be a landowner.

Romans also gives sidelights on how people ate. The diet of the slave *en voyage* was monotonous but plentiful: two pounds of rice and half a pound of pork a day. The diet of his master and his master's kin was more varied. Meat was central, with pork and ham predominating, but allowance was made also for frequent meals of beef and poultry, supplemented by eggs, bread, cakes, pies, tea, and coffee. The travelers had no fruit. The preferred alcoholic drinks seem to have been rum and madeira, with no place for gin and beer, which were the common drinks in contemporary England. Although small provision was made for supplying vegetables during the voyage, they were an important component of the diets of the planter and his slaves once he was established. The family would enjoy homegrown corn, pumpkins, peas, rice, and potatoes. The modest dwelling envisaged for the small planter Romans described as built of square cypress timber and comprising two rooms, a loft, and a corn house. No doubt he was recalling ones he had previously seen in West Florida.[41]

While Bernard Romans did what he could to make the Mississippi area of West Florida more attractive to immigrants (pointing out, for example, that if they went there they would have to take with them only four months' worth of provisions instead of the year's supply necessary for settling near Pensacola), the provincial assembly enacted legislation to facilitate and encourage settlement in the western region. It passed an act guaranteeing liberty of conscience and the free exercise of religion to settlers in the west and legalized an ad hoc body of twelve elected by the westerners themselves to rule and legislate and adjudicate for them in all minor affairs. The twelve were to choose a president with the governor's approval, and three of them would suffice for a quorum. This constitutional experiment was eventually disallowed by superior authorities, but it certainly indicates that the Pensacola establishment approved western settlement (naturally so, since many of them had invested in western land, which, with settlement, was bound to appreciate in value).[42]

It is very difficult to assess the impact of immigration upon the population of West Florida. Surviving figures for population are fragmentary and are, at best, estimates. If the odd coureur de bois and upcountry Indian trader are excepted, when the British took over West Florida in 1763, whites lived only in Pensacola and Mobile and their environs. In Spanish Pensacola all civilian residents had depended economically on the garrison, so that when the garrison evacuated the town, so did the civilians. In and around Mobile, by contrast, were numerous French families that preferred to stay put. Mobile, with 350 townspeople, was comparatively populous; in the surrounding countryside were another ninety French families, according to an official report of Lieutenant Colonel James Robertson.[43] This figure roughly accords with the numbers given by an anonymous British correspondent, probably an army officer, who wrote a lengthy description of West Florida in November 1763. Usefully he distinguished the inhabitants by color, assessing the white population at 120 and the black at 500.[44] At 620, his total was probably an underestimate but much closer to the truth than the estimate of 6,000 made by Major Robert Rogers, who was then living in London.[45]

During the succeeding period of military government and the governorship of Johnstone, the colony's population expanded rapidly. The British garrisons, initially two infantry battalions each of 500, arrived with the usual complement of camp followers. In their wake came merchants, often Americans, lured by the prospect of trading with Spanish neighbors, government officials, speculators, and smallholders. By February of 1765 the provincial population, exclusive of Indians, comprised 1,473 whites and 842 blacks.[46] A year later Governor Johnstone estimated that the civilian population numbered between 1,800 and 2,000. If the garrison were added to it, the province's population was probably close to 3,000.[47] In the years that followed, Pensacola displaced Mobile as the chief city, while new settlements sprang into being in the west. When writing his "Description of West Florida" at the beginning of 1774, Elias Durnford estimated that 2,500 whites and 600 blacks lived in the western settlements on or near the Mississippi. In the remainder of the province were 1,200 whites and 600 slaves. If these estimates are accurate, the entire population of West Florida, exclusive of Indians, on the eve of the revolution was 4,900, and the population of the gulf ports had somewhat diminished since the mid-1760s.[48] This information jibes, as far as Mobile was concerned, with a report from the Reverend William Gordon, who, in his official capacity, almost certainly kept records of births and deaths in the province. In 1774 Gordon estimated that Mobile's population comprised 330 whites and 416 blacks.[49] Much more guessing was involved in estimating the population of all areas outside the gulf ports. Durnford was one of the few writers to make even an estimate of population in the new western settlements. He wrote that the greatest concentrations were on the river Amite, where some forty families lived, and at Natchez, where a similar number lived at the time he wrote, although he was aware of almost daily new arrivals there.

J. Barton Starr has written in more detail on West Florida during the revolution than has any other historian and has attempted the extraordinarily difficult task of assessing its population growth.[50] There was no census and no record of squatters upcountry, nor of those who left West Florida, as some undoubtedly did, for the greater security of life in Spanish Louisiana or elsewhere. We have no list of births and deaths for the years of revolution, and no medical records detailing the effect of the lethal epidemics which we know periodically afflicted the colony. Starr is generally conservative in his estimates. For instance, as we have seen, on 1 April 1766 the governor gave his province's population as between 1,800 and 2,000, exclusive, it seems, of garrison.[51] Starr estimates the entire population three years later at 2,000. It is unclear why, during years that apparently should have seen vigorous growth, there was no expansion: the number of land grants in those three years—well over 400—scarcely suggests a contracting economy. Cecil Johnson was also probably conservative in estimating that, between 1774 and 1779, the years of the major Mississippi immigration boom, the population of the western part of the province merely doubled.[52] Starr not so conservatively assessed the population of the whole of the province as between 7,000 and 8,000 by 1781.[53]

Starr's figures seem a little high. A study of 284 applications made for plantation land between 24 July 1772 and 17 May 1779, the boom years for West Floridian settlement, reveals an interesting pattern. Not all the applicants were petitioning for land on family right: sometimes it yielded more acreage to apply as a demobilized veteran, and for a time during those years, family right grants were forbidden. Nevertheless, the 284 applicants who mentioned dependents had a total of 93 wives, 373 children, 16 less closely related kin, 86 indentured servants, and 1,065 slaves.[54] This summary suggests that bachelordom, as might be expected in a frontier province, was commonplace and that slaves were much preferred to white indentured servants as a labor source. Superficially too it might suggest that, since 852 whites and 1,065 blacks were listed in the applications, slaves outnumbered whites in West Florida. Although it is probable that the proportion of blacks to whites was rising in the colony's later years, for at least two reasons any such assumption is questionable. First, because all the applications were for plantation land, they were all from that segment of the population that used slaves in the greatest numbers. People who were not in West Florida primarily as planters—the townies, merchants, soldiers, sailors, Indian traders, and lesser government officials—needed few slaves or in most cases none. Second, these applicants would not include many white people who were living off the land, who had settled in remote regions and found it difficult to get to a magistrate to stake their claim to land legally. A high proportion of these squatters in the remoter regions of West Florida were people driven there by hardship from the Virginia and Carolina backcountry. They were usually poor, so much so that Governor Chester was known to use government funds to help them survive. Of course most of them were too poor to be slaveowners or even to afford the fees necessary to obtain "free" land. Poor farmers elsewhere in the province postponed obtaining title to their land for the same reason.

The people cited in these applications numbered nearly 2,000. Even if this figure excluded the truly poor, those living inaccessibly, and those who either did not want land or were already satisfied with what they had—rarities in the province once the rich western area opened up—they probably did not outnumber those in these petitions by a ratio of four to one. Three to one seems more plausible. It is also plausible that the population diminished after 1779. As chapter 7 will show, a high proportion of self-styled loyalists (on whom Starr largely relies for his high estimate) in fact went back home to New England after the Spanish conquered the region most attractive to planters in 1779. In 1780 the Spaniards went on to conquer Mobile, the center of the fur trade. Many West Floridians were thus deprived of the means of making a living. Pensacola, as the last place in the province still flying the British flag, of course increased in population in 1780 and 1781. Most of the newcomers were soldiers and sailors, now totaling 1,200, who were preparing to hold the town against expected besiegers, but there would also have been refugees unwilling or unable to leave the province.[55] Many of those who preferred not to live under Spanish rule probably had not opted for the rigors of siege life, especially if they had families, but instead followed the example of Bernard Lintot (a man of numerous family who had once been awarded 1,000 acres for proven loyalty to the British crown): they tried to find a means to return to their state of origin.[56] The courteous Spanish conqueror of the western districts had offered any settler who did not want to stay in Florida as a Spanish subject ship passage to any English port he or she chose.[57] In this regard John Fitzpatrick's comment in 1785 that "the *five* English that still remain in the country are treated with great indulgence" may be relevant on two counts.[58] It confirms the fact that the Spanish were not oppressive but that, nevertheless, there was an exodus of settlers from the Mississippi. The high point for British West Florida's population was probably the first months of the year when the Spanish took over all the province's western districts. In 1779, after sixteen years as a British colony, West Florida had a population of perhaps 6,000. As such it compared favorably with Georgia, which in 1750, at the end of eighteen years of settlement, had a population estimated at 5,200. Given peace and continued British rule, West

Florida's 6,000 would have provided a sufficient base for development comparable to that of Georgia, where the population leaped to nearly 10,000 by 1760 and then, in the following decade, more than doubled to 23,375; by 1780 it totaled over 56,000.[59]

2

Blacks in West Florida

Great Britain's province of West Florida was created partly from what had been French Louisiana and partly from Spanish Florida. Under Bourbon rule the sparsely peopled area, which consisted of little more than the settlements of Mobile and Pensacola, contained black slaves. An eyewitness who was with British troops at Pensacola when the Spanish abandoned the town in September 1763 and who was also with the force occupying Mobile on October 20 noted a difference between French and Spanish practice. Although Pensacola was left completely deserted, some 120 Frenchmen remained at Mobile with what the British observer considered the meager number of 500 black slaves.[1] This anonymous witness might have known that other Frenchmen, who stayed on after the British takeover on plantations far from the gulf ports, also owned slaves. Francis Du Planty, for example, had a plantation on Lake Pontchartrain with some white servants and 50 black slaves.[2] Another eyewitness account confirmed that the Spanish evacuated to Cuba the slaves which they had once maintained at Pensacola, when Florida became British.[3] This disparity in policy undoubtedly accounts for the comparative frequency of French slave names and the paucity of Spanish ones in the records of West Florida.

In encouraging settlement in the new colony of Georgia thirty years before, the London government had attempted to exclude slavery. No trace of this sentiment remained in the 1760s, when the ministry sought to stimulate emigration to the Floridas by offering land free except for a small annual quitrent of a halfpenny an acre. On instructions from London the governors of East and West Florida published proclamations promising any head of a household a hundred acres. An additional fifty acres was allowed for each dependent accompanying him or her, whether child or adult, man or woman, "white or negro."[4]

The Georgia experience had probably persuaded the authorities in England that slavery was essential to the economy in the American South. So, certainly, thought George Johnstone when he was angling for his governorship; the report recommending his application (probably worded by him) referred to "slaves, without which it will be impossible to raise the colony to any eminence."[5]

By enabling slaves as well as kinsfolk to qualify for free land under family right, the government undoubtedly gave the prospective immigrant a powerful incentive to go to the expense and trouble of bringing his own slaves with him. We know of several cases in which slaves were brought into West Florida purely to extend a land claim. Jeremiah Terry, a Pensacola entrepreneur, admitted to the provincial council that, to qualify for land, he brought to the province forty-five German indentured servants and black slaves, "at different times, twenty negroes."[6]

In contrast to her sister colony of East Florida, which had no slave code until 1782, the province of West Florida made detailed legal provision for controlling blacks comparatively soon after its establishment. Its elected assembly met for the first time on 3 November 1766. Eight days later a motion in the provincial council for a slave bill was approved, and a committee of three councillors, John Hannay, William Leitch, and David Hodge, was ordered to draft it. After receiving the mandatory three readings in the council, which doubled as an upper house of the legislature, the bill was passed on to the assembly for action. As "an act for the regulation and government of negroes and slaves," it became statute law on 3 January 1767.[7] Considering the constitutional convolutions it had to undergo, the slave bill was handled with a dispatch which suggests that it was an urgent matter for the whites of West Florida.

Its preamble anticipated the necessity, for the full settlement of the province, of employing "a great many Negroes"—preferably slaves rather than free blacks, it would seem, because manumission was made expensive. An owner wishing to emancipate a slave had to deposit £100 with the provincial secretary as a pledge that the freed slave would never become a burden on the colony.

Punishments for blacks were severe. Slaves might receive up to twenty lashes if found more than two miles from the town or plantation where they were employed, unless they carried a certificate from their owner. By the provisions of a separate act passed on the same day, slaves who cut adrift or stole a canoe were liable to as many as thirty-nine lashes for a first offense and death for a second.[8] Striking a white was punishable by death, even for the first offense. By contrast, the maximum punishment for an owner who neglected his or her slaves was a mild fine of forty shillings. The official responsible for runaway slaves was the provost marshal, who housed them in the town jail, fed them, and clothed them. To recover runaway slaves, an owner had to reimburse the catcher with ten shillings a head and a mileage allowance calculated at fivepence halfpenny per mile. The provost marshal himself could reclaim the various costs of advertising and keeping a runaway, either from the owner or, if none such appeared, by selling the slave.

Slaves might legally buy no alcoholic drinks, sell nothing, keep no farm animals, and carry no firearms except while under their master's control. Justices of the peace were given wide powers to disperse meetings of blacks which might disturb the peace.

Capital offenses by slaves could be dealt with swiftly. Two justices of the peace and three freeholders had to try the accused within eight days of the crime. If found guilty, they could execute the slave as soon as they saw fit. Free Indians, mulattoes, mustees, and blacks could give evidence in all cases where slaves were on trial.[9]

The code allowed justices who were trying blacks to temper their sentences if mitigating circumstances existed, but for certain offenses—arson, slave stealing, and poisoning—the death penalty was mandatory. Stiff penalties were prescribed for masters unwilling to allow their own slaves to appear as witnesses at the trials of other slaves and also for masters trying to protect their slaves from the courts. The provost marshal administered punishments. He was paid ten shillings for a public whipping, two shillings and fourpence for a common whipping, and one pound sterling for an execution. He thus had a vested interest in severe penalties.

A constable who withheld information about illegal activities involving blacks could be fined. Much more heavily fined were freeholders, justices of the peace, and provost marshals who failed to perform the duties assigned to them. Although the act specified that slaves were property—"to all intents, constructions and purposes, chattels personal in the hands of their owners"— masters could not necessarily do what they liked with their property. Killing one's slave was punishable by a fine of £100 for the first offense and by death for a subsequent offense. If the colonial judicial authority executed a slave for criminal activity, it had to pay a reasonable compensation to the master.

In summary it may be said that this first slave statute of 1767 sought to create and perpetuate an inflexible system in which all those who were not white—color was the "badge of slavery," read the act—had an important but restricted and subordinate place in society. To make the system work as designed, responsibilities were spread widely among the white population from the council down to the large freeholder class, and those who evaded their responsibilities were penalized. Certain officials, particularly the provost marshal, were well rewarded for their role in operating the slave system. Surprisingly perhaps, the code was not copied directly from that of Georgia or South Carolina but nevertheless contained, as did the slave laws of those colonies, a hinted fear of slave rebellion.[10]

Members of the lower house of the West Florida legislature soon felt the need for a new slave act, and on 2 June 1767 they presented their superiors with a bill "for the order and government of slaves" that was designed to replace the similar act passed only six months before.[11] The council approved it after minor amendment, and the acting governor signed it on 6 June.

Why the original act was thought unsatisfactory remains unknown, although it was not because it was too harsh. Between January and June, certain practices by blacks like drumming and feasting, which are mentioned in the later act alone, had perhaps become a widespread and arguably dangerous irritant. Innovations in the second bill also suggest that, in the same interval of six months, there may have been a rise in theft attributed to blacks. Alternatively, since the original bill had been drafted by a committee of three townies, William Leitch, the chief justice of the colony, John Hannay, the acting provost marshal, and David Hodge, a merchant, it is possible that the assembly's planters, men of large acreage and numerous slaves like Alexander Moore and William Aird thought that a tougher slave code than that devised by the urban trio was called for, especially as the Floridas, unlike other southern colonies, had no militia to help control slaves.[12] West Florida's second slave act was certainly much stricter than its first.

It specified higher rewards for those who captured runaways. The original act mentioned death as a fit punishment for slaves guilty of grave crimes; the new act specified dismemberment as another possibility. In addition it doubled from ten to twenty pounds the penalty for those freeholders who neglected their duty in regulating slaves, and it more than doubled the penalty for anyone trying to prevent a slave guilty of a capital crime from receiving his legal deserts. New offenses were defined; for example, a slave who ran away and avoided recapture for more than six months was considered ipso facto to be guilty of rebellion and would be liable to transportation. For a slave this sentence meant being handed over to a ship's captain with instructions that he be sold at a distant port to anyone who would buy him. According to W. Robert Higgins, the prospect of being sent to an unknown destination for purchase by an unknown master was a significant psychological deterrent: the slave did not see it as a chance for a better life.[13] After transportation, if he dared to return to West Florida, he would be hanged. Under the terms of the new act, slaves might not play games of chance for gain. If they did, they could be whipped through the streets. There was new emphasis too on theft. Owners of slaves had a legal obligation to cooperate in the search for stolen goods in slave cabins, while buying stolen goods from a slave was defined as a felony. The previous penalties for killing slaves still stood, but now slave murder was legally excused if the slave was not on his master's property.

In the new bill the few pleasures open to a slave were curtailed. Meeting in numbers on a Sunday or holidays was now forbidden. So was making a noise, especially drumming, and a ban was now declared on the "bad practice" of slaves feasting together at night. A constable was to earn a reward of three ryals, or three-eighths of a dollar, whenever he found more than six slaves together after nine in the evening. Another new feature of the bill was an explicit denial that a slave might qualify for freedom by becoming a Christian.

The new act was identical in intent with the one it was designed to replace, although its provisions were more vindictively (though not always more precisely) expressed. Eventually the privy council of Great Britain vetoed it, but only after an unusually great delay. The first official suggestion that it should be disallowed came from a newly appointed governor, Peter Chester, on Christmas Day 1770, three and a half years after the bill had received a predecessor's assent.[14] Chester's sole objection to the act was that it specified the swift execution of a slave found guilty of a capital crime but allowed no opportunity for the king to exercise clemency. Richard "Omniscient" Jackson, who advised the British government on colonial legislation, saw nothing wrong with the act and, on careful reconsideration after an interval of months, still found it acceptable.[15] He scouted Chester's argument against it, but the London government ignored his advice on the second West Florida slave act, and the king in council invalidated it on 15 January 1772.[16] So slow was administration and transatlantic communication that not until 4 November 1772 did Governor Chester ask the advice of his council following the home government's disallowance of the 1767 "act for the order and government of slaves." The chief justice and the majority of the board were clear in their opinion that the invalidation of the second 1767 act meant that the first, repealed act on the subject of slaves was "now revived in full force."[17] The invalidated act had been enforced as if legal for over five years. One may only hope that, if the second 1767 act had contained provisions that led to flagrant abuses of contemporary concepts of justice, the wheels of administration would have turned more swiftly.

Most evidence suggests that the slave acts were enforced with severity. An exception occurred in 1770, when one of Major Robert Farmar's slaves, one Drummer, was convicted of burglary and felony on 24 August. Under the terms of the slave act, he was liable to the death penalty. He was in jail in

Mobile awaiting execution when, on 23 September, Governor Chester, who seems to have been preoccupied with clemency that year, pardoned him upon condition that he leave Florida within a month and never return.[18] The Indian traders, too, ignored the slave acts, as they did all laws that harmed their profits. Bold men, the traders roamed the woods and lived in Indian villages and were not easily disciplined. Although both slave acts prohibited sales of alcoholic drinks to blacks,[19] the superintendent of southern Indians, John Stuart, recommended, as late as 29 October 1778, that the West Florida legislature take effective action against traders who sold or bartered spirits to blacks and that it stop blacks from selling rum to the Indians—which suggests that blacks were defying the law by both drinking and trading in spirits.[20]

Conditions in West Florida made it tempting for slaves to try to escape. The slave code was harsh, but its enforcement was difficult. The province was huge, and its population tiny. Foreign settlements, both on the mainland and on Caribbean islands, were much closer to West Florida than were most other British colonies in America; there, a runaway could hope to be safe, at least from British slave catchers.

It was natural that representatives of the imperial government should cooperate in stopping leaks in the tight slave system constructed by provincial legislators. At a congress in 1767 in Augusta, Georgia, John Stuart obtained Creek and Cherokee agreement to surrender any runaway blacks sheltering in their lands.[21]

Since there was no newspaper in West Florida, information about runaways from that province has to be gleaned from the papers of neighboring colonies. We learn from the *Georgia Gazette* of 20 June 1765 of Prince, a tall, well-made, sensible carpenter and cooper. He had deserted his owner, John Watts of Pensacola, who thought that, being an "artful, cunning fellow," Prince might have tried to pass himself off as a freedman and had perhaps changed the blue clothes he was wearing when he escaped. Parenthetically, blue was a common color for slave clothing; in the short Floridian winter, over garments of blue Limburg cloth a slave might wear a hooded cape made from a blanket of another color.[22] For a slave as valuable as Prince, the offered reward of twenty dollars and reasonable expenses seems parsimonious. Even less, twenty shillings a head, was offered for the recovery of four blacks who deserted from the plantation of Robert Bradley of Pensacola. Three of them had the grandiloquent classical names of Neptune, Bacchus, and Apollo. They were "new negroes" (that is, recently imported) and spoke no English. The fourth, Limerick, was presumed to have led the others astray, possibly to seek refuge among the Creek Indians. It was probably not the first time that Limerick had defied authority, because he was described as being much marked "by severe whipping." Maybe it was one of this quartet who was picked up over a year later in the Creek nation. He was described as a "new negroe fellow . . . of the Bumbo country." This runaway had learned the Creek language, in which he told his captors that he came from Pensacola, had been with the Indians for twelve months, and used to fetch wood for his master, a "man with a big belly."[23]

An alternative to trusting in Indian hospitality was to pretend to be free and to seek employment on a vessel leaving the colony. Such was the choice of James, who worked for an East Floridian master. Having lived in Virginia and Carolina, perhaps James had a better understanding of the wide world than had many slaves. In his case the extraordinarily high reward of fifty pounds was offered for information about any white person who had helped his escape. It would seem certain that, such was the small size of the ports of the British Floridas, no runaway could possibly have escaped, as James did, without assistance from a white protector.

Following a tradition established before 1763, when slaves from Georgia ran away to St. Augustine, in the often-justified hope that the Spanish authorities would recognize them as freedmen, slaves in later decades sometimes fled to the British Floridas as well as from them. A slave who, though not a "new Negroe," had an African name, Cudjoe, left his South Carolina plantation in 1771. His master thought he had headed for West Florida, with which he was already acquainted. As well as a reward of ten pounds for his recovery, the master, Peter Simons, offered twenty pounds for information leading to the conviction of any white man who had sheltered him.[24] Another slave who sought refuge in West Florida was advertised as having originated in North Carolina: he lived for 4½ years with the Indians as a runaway before, tiring of the life, he went to Pensacola still wearing the iron ring, "uncommonly well fixed," which was on his leg when he escaped. The deputy provost marshal of West Florida, Alexander McCullagh, advertised that, if the man were not reclaimed within a year, he would be sold at public auction to recover public expenses.[25]

It is worth noting that, whatever the rivalries dividing their governments might be, slaveowners of all nationalities perceived a common interest in cooperating in the return of runaways, no matter how far fled. The trouble and expense of recapture and return were of minor importance. Of major significance was the demonstration to potential runaway slaves of the futility of attempts at flight. John Fitzpatrick at Manchac sought the help of the Mobile trader Peter Swanson in catching two blacks who had fled from the plantation of a Monsieur Trenonay at Pointe Coupée in Spanish Louisiana and who had been seen at the plantation of John Murray in West Florida. If caught, the two were to be sent back to New Orleans, regardless of cost. Fitzpatrick also asked Swanson's assistance in recovering for a French officer then in the distant Illinois country a mulatto slave who had left the Frenchman to take refuge among the Chickasaw. Again the full cost of returning him to New Orleans would willingly be repaid. Swanson expected and received similar cooperation from Fitzpatrick. Those unfortunate enough to be recaptured were invariably shackled, though not always effectively.[26]

The diary of Fitzpatrick's neighbor William Dunbar is the best existing source of information about slave life in West Florida. When Dunbar began writing he had fourteen blacks. Seven were women, of whom three were house servants. The remaining eleven worked in the fields and woods and were supplemented by a less useful, pitiable batch of "new" blacks, twenty-three in number, whose primary reason for existence as far as Dunbar was concerned was as merchandise to be sold at a profit.

Dunbar employed a white overseer but did not remove himself from contact with his slaves. He knew exactly what each was doing every day and planned their days with consideration to the weather, the tasks needed to maximize profits and food production, and the health of his slaves. Essentially he considered them as ignorant children who ought to be grateful for the food and clothes he provided, who must be denied access to liquor, and who for their own good should be physically punished if they broke the plantation rules. For such transgressions as stealing liquor, fighting, and running away, Dunbar prescribed whipping, chaining to a heavy log, and confinement in a "bastile," presumably a dark cellar or small shed.

His slaves' work was hard and long and included tree felling, swamp clearing, roadmaking, cane cutting, rice cleaning, and stavemaking. The women were not exempt from work in the plantation sawpit. Every slave received a jacket and a pair of shoes, and in addition the men were given pants and the women petticoats. Although food was varied and plentiful, sickness was common for the slaves and their master on the low-lying Dunbar plantation.

Dunbar was highly intelligent and of course knew that it was in his own interest to have healthy, well-fed slaves. His diary leaves the impression that, in his relations with his slaves, economy was more important than human concern. When describing the death of Cato, the best worker on the plantation, Dunbar's final comment was that he "would have fetched at market a hundred pounds of sterling."[27]

Dunbar's new blacks may have come from Africa, but West Florida's slave population came from many sources. Until the Spanish established their rule in New Orleans in 1768, French Louisiana was a source of slaves for West

Florida. There were slaves for sale at Pointe Coupée and probably the capital city too in 1764.[28] A British detachment ordered to clear the Iberville managed to hire fifty black slaves for the job in 1764. Since the western part of British West Florida had the scantiest population at the time, they almost certainly came from Louisiana. By the following year, the economy of the foreign colony across the Mississippi was in a slump, and blacks were harder to find. Captain Harry Gordon would comment on their scarcity there in 1766.[29] British entrepreneurs sold as well as bought slaves in Louisiana. To do so was against Spanish law, and one of several restrictions imposed on his province by the new Spanish governor, Antonio de Ulloa, which ultimately led to a revolt of the citizens of New Orleans, was an embargo on all trade in blacks with English colonies.[30]

Sometimes Louisianan slaves came into West Florida, not as the result of any commercial transaction, but because their owners migrated to the British colony. There were at least thirty such migrant families, of which we know at least one, the Monsantos, to have been slaveowners.[31] Five other Louisianans had offered to emigrate to West Florida in 1769 if they could be promised land grants. They were François Broutin, Albert Bonne, Stephen Arlic, James Chapiron, and André Carrière. The five pledged to bring ninety-seven slaves with them, but the West Florida council did not accept their offer at once, and it is unsure that they ever arrived.[32]

The Spanish ban was difficult to enforce, partly because the Paris treaty of 1763 had guaranteed the free navigation of the Mississippi to the British. Sporadically the Spanish authorities made life awkward for British traders. In September 1769 the newly arrived military governor, General Allessandro O'Reilly, insisted that they might not set foot on Spanish soil, to the annoyance of Captain Richard Nicholls, who had a cargo of thirty slaves in his brig *Sea Flower.*[33] Such a restriction could have made it all but impossible to bring a vessel up to New Orleans, since negotiating the sharp turns of the Mississippi in the days of sail meant warping. There were several ways of performing this tedious task. In eighteenth-century Louisiana it meant helping a boat negotiate bends either by placing a crew onshore to haul on a warp (or light hawser) attached to the vessel or by using a capstan on board, attached to a fixed object, such as a tree, on the riverbank. Either method necessitated landfall, and both were slow.

In general, Spanish governors of Louisiana were not rigid. Even the strictest of them, O'Reilly, according to Nicholls's account, was prepared to bend regulations in his case, to the extent of letting him sell two of his slaves to a Frenchman in New Orleans. But the comparatively tough regime of O'Reilly was short lived, and his predecessors and successors were readier to see that the economic welfare of their colony depended on winking at the rules. The citizens of the Spanish province needed from their neighbors American flour, European manufactures, and black slaves, which were in particular demand, since there had been no considerable importation since 1743.[34] As West Florida developed and the center of commercial gravity shifted westward, Natchez became important as a source of staples, and Manchac on the Mississippi above New Orleans became the heart of a contraband trade conducted by means of "floating warehouses" which could legally anchor in the Mississippi off New Orleans and Pointe Coupée. When in 1776 and 1777 John Fitzpatrick, the Manchac trader, and his friend, the Mississippi planter William Dunbar, sought to dispose of slaves, they looked to Pointe Coupée, and Dunbar and the planter Alexander Ross bought slaves in some numbers from New Orleans.[35]

In 1777 the easygoing regime of Governor Unzaga ended. Although the British did not at first realize it, his successor, Bernardo de Gálvez, was more forceful. Eleven floating warehouses, two of them American and the rest British worth $70,000, were suddenly seized. British merchants arrested (besides three others, difficult to identify because of the curious way in which the Spanish transliterated their names) included John Waugh, John Campbell, the Ross brothers, and the partnership of Patrick Morgan and James Mather.[36] In the Spanish takeover Morgan and Mather lost a brig holding twenty-two blacks; too many for crew, they must have been cargo. Surprisingly this incident did not abruptly end Anglo-Spanish slave trading. In 1778 Fitzpatrick wrote without comment—as though it were the normal thing to do—that he was putting blacks up for sale at Pointe Coupée.[37]

Factors nevertheless were already at work which produced a steep decline in the trade. One was the hostile incursion into West Florida of the American, James Willing, in the first months of 1778. As a result Fitzpatrick of Manchac lost $1,300 in property, and Robert and David Ross lost the schooner *Dispatch* containing a cargo of flour and fifty slaves.[38] Both merchants survived these blows, but others may have been driven out of business. Willing confiscated a total of 100 slaves, which he promptly sold, although Gálvez compelled him to pay compensation to several owners whom he had robbed, including the Rosses.

Much more important than American depredations in diminishing the illicit Anglo-Spanish trade were new Spanish regulations introduced late in 1777 making legitimate trade more attractive to Louisianans: they were permitted to trade with Yucatan, the duty on exports was reduced to a nominal 2 percent, and, most important of all for the subject of blacks, the French were again allowed to send their slaves from Guinea to Louisiana.[39]

Finally, of course, the outbreak of hostilities between Britain and Spain in
1779 brought Spanish conquest of West Floridian settlements on the Mississippi and an end to peaceful Anglo-Spanish activities. Although precise
figures do not exist, it seems evident that, although Louisiana temporarily
and on a small scale supplied slaves to West Florida, over a longer period and
in greater quantity, Louisiana imported slaves from West Floridian traders.
Louisiana had never been a main source of supply to West Florida. From the
time of first British occupation, the British colony had also imported slaves
from elsewhere, chiefly from other British colonies.

Soon after the establishment of West Florida, as we have seen, settlers
found the province's western lands more attractive than the gulf coast. Shippers saw no reason to route their slave cargoes by way of Mobile and
Pensacola. Had the Iberville cut been made, shortening the distance to the
sea from Natchez and Manchac, then Mobile and Pensacola would surely
have benefited. The cut was never made, and far from replacing New Orleans,
the gulf ports largely fulfilled Lieutenant George Phyn's prophecy that they
would remain inconsiderable, a "receptacle for men of broken fortunes."[40]
Slave dealers who bypassed them included a London merchant, Edward
Codrington, who regularly sent cargoes of goods, both human and otherwise, directly from Jamaica to Manchac and from Charleston to Manchac.
They were exchanged for consignments of skins and indigo, which went
directly or indirectly to London. Codrington failed to include Mobile and
Pensacola on the itinerary, probably for fiscal considerations. No customs
official lived at Manchac, and it was therefore easy to evade duties. And if,
on the way to Manchac, Spanish or French colonists tried to buy part of the
cargo, again no duties would be paid, since any such transaction was illegal.

Jacob Blackwell, collector of the customs at Mobile, had a close business
connection with Codrington. He was not irresponsible about customs collection, but he may have been willing to ignore such evasion by his patron.[41]
Nancy and Molly, the brigantine in which Codrington shipped some of his
cargoes, was the property of the Philadelphia firm of Thomas Willing and
Robert Morris.[42] Commanded by Robert Bethel, she voyaged between Philadelphia and the Mississippi in 1772, 1773, and 1774.[43] Trade between revolutionary Philadelphia and loyal West Florida ceased in 1775, and so *Nancy
and Molly* was sold for £1,200 to Oliver Pollock, a New Orleans merchant,
in July 1775. Similar in function to *Nancy and Molly* was the schooner *Rose,*
carrying a cargo of slaves from Charleston to the Mississippi, which a Richard
Skinner bequeathed in his will to a brother.[44]

In the prewar years another firm which traded in slaves on the Mississippi
was based in Jamaica. The partners, John Bradley and Joseph Harrison,
described themselves as Kingston merchants, but Bradley spent much time

on the Mississippi. In 1770 they owed the London merchants Walker and Dawson £7,000 and as security for repayment pledged the snow *Africa* and its cargo of eleven slaves, which were destined for sale when she reached the Mississippi.[45]

The firm of Baynton and Wharton of Philadelphia also did much business on the Mississippi without troubling the authorities on the gulf coast. The ignorance of the authorities is the more understandable because, although some of their blacks were certainly brought from Jamaica, others were apparently sent from Pennsylvania to Florida by the way of the Ohio.[46] This approach to West Florida from "the blind side" was commonplace in the 1770s. As late as 1778 Pierre Langlois traveled with a bargeful of stock, including two slaves from Kaskaskia in the Illinois country, in order to sell his wares lower down the Mississippi.[47]

For the gulf ports the West Indies were conveniently located, and Jamaica was a main source of slaves in the 1760s. Pensacola received at least 6 from that island in 1765, 59 in 1766, 120 in 1767, and 23 in 1769.[48] Sometimes slaves were transferred rather than purchased from Jamaica in order to enhance a claim for free land or for other reasons. When Walter Hood petitioned for a large 2,000-acre tract on the Mississippi in 1768, he justified his claim by citing the expected arrival in West Florida of a brother with slaves from Jamaica. In 1769 Patrick Morgan applied for a similar grant, explaining that he was awaiting the imminent arrival of fifteen blacks from the same island,[49] and when William Dunbar obtained land on the Mississippi in 1773, it was to Jamaica again that he voyaged to buy slaves to work his new land.[50] John Priest too emigrated to West Florida and brought slaves with him from Jamaica, selling some of them at Manchac.[51] War in the continental colonies made it difficult for planters to feed their slaves in Britain's island colonies. Alexander Murray and William Prescott sought to maintain their prosperity by transferring their "families" from Jamaica to West Florida.[52] Another wartime migrant to West Florida was Samuel Steer. Originally a trader based in Georgia and Jamaica, he acquired a plantation on the Mississippi in 1777 and placed on it twenty Jamaican slaves fully equipped with winter and summer clothes, utensils, and implements.[53]

Thus did the fortune of war increase the black population of the province, but the traditional importation also continued. In 1778 the marauder Joseph Calvert, acting the American patriot and equipped, according to Governor Chester, by James Willing and Oliver Pollock, captured the previously mentioned schooner *Dispatch* when she was acting as a slaver from Jamaica.[54]

Other islands supplied West Florida with black and white war refugees. From St. Vincent came William Walker and Levi Porter with forty-three slaves between them.[55] Robert Taitt settled 550 acres on the Escambia with his nine blacks after fleeing from Grenada. He alleged that Oliver Pollock had taken a vessel of his containing nearly 200 slaves.[56] Once established in Florida, Taitt was politically active and represented the Pensacola district in the assembly in 1778.[57] It seems likely that he continued trading in slaves as a sideline.

Britain's mainland colonies were another important source of supply. A cargo of slaves imported into Pensacola in 1767 evidently came by way of Georgia. The master of the vessel was Captain John McCoy, who made several trips in the *Fanny* between Savannah and West Florida, where he had become the owner of a building plot in 1765.[58] Often, it seems, Savannah was one of a chain of markets through which slaves passed on their way to Florida. For example, the schooner *Margaret* carried a total of fourteen blacks from Savannah to East Florida in May of 1767. Seven of them were "new negroes" imported earlier in the month from St. Vincent to Savannah, and five were new and two were seasoned blacks imported from the Grenadines.[59] Many slaves arrived in Florida, as often they did from the islands, not as part of marketable assignments, but in company with their emigrating masters, particularly in the 1770s. From the Carolinas an overland emigrant group of seventy-nine poor whites arrived with eighteen slaves between them, who wanted to settle near Natchez.[60] A considerable number came from the Virginia backcountry in the early 1770s. On 5 May 1773 Rufus Putnam, while exploring the most northerly section of West Florida, met Jacob Winfree, who had arrived from Virginia that very month, bringing fifteen or sixteen slaves with him. They were in company with about a hundred Virginian men and women of varied color.[61]

Naturally the outbreak of revolution ended the slave trade between South Carolina and the Floridas. It did not end black immigration. As the war against the king flourished, many of his supporters looked for resettlement in loyal West Florida, especially when it became known that on 11 November 1775 its governor had issued a proclamation offering not merely asylum but also free land, beyond what could be claimed by family right, to immigrants who had suffered loss through loyalty to the crown. In contrast to prewar groups from poor backcountry areas, those who arrived during the war were often from prosperous districts and themselves still affluent enough to bring numerous slaves with them. Usually they sought plantations in the western part of the province. The Lintot family, which brought seven blacks with them in 1776, came from Connecticut.[62] In the same year the Bassets brought fourteen slaves and William Webb five from Georgia. In 1777, William Marshall Junior arrived with twenty blacks, and David Holmes with twenty-two. They had traveled from South Carolina.[63] In 1778 a group of about thirty-five immigrants under Colonel Tacitus Gaillard and Dr. Benjamin Farrar, also from South Carolina, arrived in Natchez with between 400 and 500 slaves.

This extremely large group was armed and had made its way via the Ohio and Mississippi rivers from Pittsburgh in five big bateaux equipped with cannon. At first its appearance at Natchez had excited alarm. Residents feared, understandably, that something like the Willing raid was again afoot. The newcomers soon made it clear that their intentions were peaceful: they wanted to settle in West Florida. For several reasons suspicion lingered. One was the unwillingness of their leader to take an oath of allegiance to the British crown until he had seen Governor Chester at Pensacola. Another was their background. Tacitus Gaillard had been a member of a rebel provincial congress. Benjamin Farrar, a former deputy surveyor in South Carolina, had been a Son of Liberty. Alexander McIntosh, the Indian commissary who had to entertain them at Natchez, thought it prudent not to insist that they take the allegiance oath there. Governor Chester had heard that the pair were supporters of the popular party in the Carolinas up to the time of the declaration of independence, after which they had turned decisively against the American cause, but he remained uncertain about their ultimate intentions. John Stuart shrewdly suspected that their aim was to live under the Spanish flag.[64] Had such a contingent remained on British soil, it is interesting to speculate on how the history of West Florida might have been changed. Effective resistance to the Spanish conquerors of Natchez in 1779 might conceivably have resulted if Gaillard and Farrar's conversion to loyalism was genuine.

From North Carolina in 1780 came James Baird with impressive credentials and twelve slaves. He was granted nearly 2,000 acres. Other loyalists with numerous slaves formed a Virginian enclave on the Pearl River, while a pair of Pennsylvanians established themselves in West Florida with nineteen black slaves.[65]

In addition to slaves imported from more distant colonies, it is very probable that African-born slaves were also brought from their native continent to West Florida via St. Augustine. The entrepreneur who supplied East Florida with new slaves was Richard Oswald, whose position and interests made it easy and desirable for him to do so. He was a London merchant with estates in the West Indies and a thriving plantation in East Florida. He had the ear of government in both London and St. Augustine, and at his initiative a first shipment of 70 slaves from Africa was disembarked at St. Augustine from the brigantine *Augustine Packet,* commanded by Richard Savery, in 1767. A further 150 slaves disembarked in St. Augustine in 1769,[66] and Governor Grant reported with great satisfaction the arrival of a third Oswald cargo of "very fine slaves" in January 1770.[67] Again under the auspices of Oswald, 120 slaves from Bance Island (Sierra Leone) were disembarked from the snow *Charlotte* by its master James Toth at the end of January 1771.[68] Three years later, the acting governor of West Florida mentioned that some forty families from the Carolinas and East Florida had settled on the Amite River. Most of them brought blacks with them, he wrote, and it seems likely that they would have included some brought to America by Oswald.[69]

There are few clear instances of slaves who were imported directly from Africa, and it is impossible to find any with a satisfactory amount of supporting detail, but some existing evidence does substantiate direct importation of new negroes into West Florida. The fragmentary tax records of West Florida show that twenty-one "new negroes" were legally imported in 1767, one by Captain [William?] Moore, two by Captain Jabez Johnstone, ten by Captain [John?] Bradley, and eight by unnamed others. In 1768 nineteen were imported, one by a Captain Samson and all the rest by Valens Stephen Comyn, a prominent Pensacola merchant and provincial assembly member.[70] Comyn's blacks, at least, may have come directly from Africa.

Comyn was the factor for his father, Thomas Comyn, who described himself as a London merchant and alleged that the West Florida branch of his business was more heavily funded, with a capital of £15,000, than any of its rivals. When applying for a grant of 10,000 acres within the province in the winter of 1768–1769, Thomas Comyn stressed, as evidence of his intention to settle in West Florida, that he had already sent "a considerable cargo of negroes" to Pensacola in order to go some way to supply the labor force for his intended plantation.[71]

Oliver Pollock, a New Orleans merchant, describing his activities in the years prior to the revolution, wrote that he was supplied "with dry goods from London, *Negroes from Africa* and flour from Philadelphia."[72]

It seems that many British merchants involved in the African trade had, like Thomas Comyn, a stake in West Florida. The British government's decision in 1768 to evacuate most of its troops from that colony dismayed all with interests in it. Twenty-three London commercial firms signed a petition protesting the decision, and a similar memorial from the headquarters of the British slave traffic, Liverpool, was signed by forty-one merchants "trading to West Florida."[73] The strength of the Liverpudlian reaction may seem surprising. But there was reason for concern: although the carriage of slaves from Africa was reaching unprecedented heights at this period, signs already existed of a diminished, even saturated, demand for them in the older American colonies. Complaints of oversupply there were followed in 1769 by an agreement of the Virginia house of burgesses to import no more slaves and, in 1774, by a similar agreement to which all members of the first Continental Congress except Georgia assented.[74] The Floridas were a necessary outlet for slaves to compensate for failing demand elsewhere. Liverpool, which by the 1760s had outstripped Bristol and London and quite dominated the colonial trade in slaves, had thus a particular interest in urging the British government to supply adequate military protection for West Florida.

Most, perhaps all, of those who signed the memorial for Liverpool were engaged in slave trading. Signatories Charles Goore, Benjamin Heywood, Henry Hardwar, William Higginston, Thomas Mears, John Strong, James Carruthers, Thomas Midgley,[75] Peter Leay, Miles Barber, Charles Cooke, Francis Ingram, and Thomas Hodgson undoubtedly were.[76] They referred to the advantageous position of West Florida, which had given them reason to look for "a very considerable extention [*sic*] of the African trade in the sale of Negroes for the cultivation of the land and to the Spaniards throughout the whole bay of Mexico."[77] In short, West Florida would be not merely the ultimate destination for their slaves but also a suitably placed market for Spanish buyers, from Campeche or Veracruz perhaps, who would find it easier to sail to Mobile or Pensacola for their slaves than to make the tedious voyage up the Mississippi to Spanish New Orleans.

West Florida was indeed a place from which slaves were reexported, which is further suggested by the laws of the West Florida assembly, which, from the first act imposing a duty on imported blacks, exempted those who were brought to the colony for the purpose of reexport within three months, provided that the importer posted a bond of £500 to ensure that he fulfilled his intention. This provision was repeated in the similar acts of 1768 and 1770, except that, instead of £500, the bond was set at twice the value of the slaves to be reexported. In the last act ever passed by the West Florida assembly, a further modification was introduced: the period within which a slave had to be reexported to be duty free was extended from three to six months.[78] The repeated inclusion of a reexport clause, particularly one that was twice modified, strongly suggests that the reexportation of slaves was not a mere hope but actual practice.

There are few precise descriptions of West Floridian slaves, but a detailed inventory has survived of the possessions of John Fitzpatrick, which was compiled at the time of his death. Although they were listed ten years after British rule ended in West Florida, six of his eleven slaves were middle-aged or older and likely to have been with him since the British period. All six were new Africans, of Senegalese, Dunbar, Temene, Papa, or Coranco nationality; the other five were colonial-born offspring of these new Africans.[79]

Of course Fitzpatrick might have bought them in Spanish Louisiana, but it is possible that he preferred to buy them from one of his many British associates in Pensacola or Mobile, with whom he maintained a much more frequent correspondence than with associates in Louisiana.

In summary, the sources of slaves in West Florida were amazingly diverse. Although the paucity of official import lists is frustrating, the slaves of West Florida who were not already there when the British first arrived came from Spanish Louisiana, Jamaica, Charleston, East Florida, the Windward Islands, West Africa, and possibly elsewhere.

Most blacks in West Florida were employed to perform the heavier work on the plantations: digging, weeding, hoeing, fence mending, stavemaking, harvesting, and tending animals. Others were domestics, with duties in the households of their masters and mistresses. Yet others worked in the towns, in the infant boatbuilding industry, or as carpenters, coopers, brickmakers, and sawyers. A number of them obtained their freedom, including Phyllis, a personal servant of Governor Johnstone who, for a nominal sum when Johnstone left his province, was passed on to his friend and secretary who stayed behind. When he too quit the colony, James Primrose Thompson emancipated Phyllis.[80]

Not all blacks were employed by private owners. The armed forces used many blacks from the early days of the colony's foundation in the interval of military government. Their use was approved by the commander in chief of British forces in North America in a letter dominated by economic considerations:

> Colonel Tayler acquainted me of the inability of the soldiers thro sickness to cut their wood agreeable to the regulation, that the contracts for the supply of wood were extravagant, and the cheapest method to effect that necessary service was to purchase negroes, by which he hoped to bring the price of wood within the stated allowance. Colonel Tayler seems to have been the only person who has used any endeavours to reduce the exorbitant prices demanded for every article in West Florida.[81]

Some blacks were used as sailors during early exploratory expeditions upriver[82] and again during the revolution, when, although numbers remain vague, blacks served aboard the *Christiana* sloop and the *Pontchartrain* galley and doubtless other vessels in the service of the crown.[83] Of more importance in wartime was their service on land. Adam Chrystie put uniforms on the backs and weapons in the hands of twenty-two of his slaves at the time of Willing's raid.[84]

In 1780 blacks labored on and defended the fortifications of Fort Charlotte in Mobile. Lieutenant Governor Elias Durnford in vain asked the military authorities to provide "insurance" against any harm that might befall them by guaranteeing compensation to owners.[85] It turned out that the Spanish conquerors kept all blacks found within the fort, to the chagrin of Mobile merchants who in 1798 were still trying to obtain some redress from the British Parliament.[86]

Perhaps grudgingly, Pensacolans too lent blacks to help prepare their town for a siege. Among them were Chief Justice William Clifton, James Bruce, David Hodge, John Stephenson, and David Baird.[87] William Johnstone was perhaps a more zealous loyalist. An artillery captain and a substantial planter, Johnstone leased thirty slaves to the ordnance board at Pensacola for the duration of the siege for the nominal sum of one ryal, or one-eighth of a dollar, a day.[88] During the siege itself, on 30 March 1781, Captain Johnstone came to the support of the Maryland Loyalists as they fought indecisively with the Spanish outside the Pensacola fort. He brought with him two fieldpieces, a howitzer, and fifty blacks. This intervention turned the tide of battle, and the Spaniards had to retire. A few days later, we learn from a siege diary, five blacks captured a Spaniard at Gull Point.[89] Putting muskets in the hands of blacks must have been a controversial decision, although it was not without precedent in the history of the southern colonies. The expedient did not save Pensacola from conquest. As at Mobile, the surrender terms specified that "all negroes, *whether slave or free* who have taken up arms belonging either to his Britannic Majesty or to individuals of the fort *shall be the property* of His Catholic Majesty."[90]

The expedient of arming slaves would doubtless not have been used if the population of West Florida had been larger. The small population was a source of several ills, including settlement in unauthorized areas by maverick settlers who could be fairly sure that they would not be ejected. They were, however, noticed. The deputy superintendent for Indian affairs in the southern department reported that one Bubbie and the better-known James Colbert had led a party of eighteen trespassing traders to hunt in the grounds reserved for the Creeks. Colbert was making no ephemeral encroachment on Indian land. He had established there a permanent plantation with cattle, slaves, and an overseer.[91] Such squatters had every reason to be inconspicuous, one reason being to avoid paying for a trader's license. They lived in remote areas and sometimes owned slaves. In another category was the talented chief of the Upper Creeks, Alexander McGillivray; in 1776 a French traveler found him living in the style of a substantial white planter at Little Tallassee on the Coosa River, where he maintained sixty black slaves.[92]

The latter half of the eighteenth century, which included the entire history of British West Florida, was the heyday of the Enlightenment. Many of the thinkers associated with this intellectual movement considered slavery indefensible. The Baron de Montesquieu in *L'esprit des lois* (1748), Francis Hutchinson in his *System of Moral Philosophy* (1745), Adam Smith in *The Theory of Moral Sentiments* (1759), George Wallace in his *System of the Principles of the Laws of Scotland* (1760), Rousseau in *Le contrat social* (1762), Blackstone in his *Commentaries on the Laws of England* (1765–1769), Adam Ferguson in his *Institutes of Moral Philosophy* (1769), the Abbé Raynal in his *Histoire philosophique* (1770), and Edmund Burke in his plan to end colonial slavery in 1780 all vilified the institution.[93] These thinkers, of whom the most influential was Montesquieu, were widely read in Britain. Half of them were Scots, and it might be noted that a large proportion of West Floridians were also Scots.

Other late eighteenth-century movements of great significance were evangelicalism and Methodism, both of which condemned slavery on theological rather than the rational grounds of the Enlightenment philosophers. The Quakers too on both sides of the Atlantic were extremely active in the eighteenth century, but most particularly in the 1760s and 1770s, in acting and arguing against slavery. We must ask what effect these various movements had on slavery in West Florida.

Of the few existing examples of theoretical writing on slavery in West Florida, only one favored emancipation. The writer, identifiable only by his initials J. P., had evidently never been to Florida; his inexperience led him into such follies as suggesting that July (the worst time for the health of unseasoned newcomers) was the ideal time of the year for disembarking new immigrants. In 1772 he wrote a detailed plan for a large settlement in West Florida centered on Manchac. In addition to 2,000 white males, he advocated the importation there of 1,000 slaves from the coast of Guinea. He gave his reasons for this supplement. He believed, for one, that "white men from a northern climate cannot stand the sun in hard labour and drudgery of plantation work in the open field in this hot country." He argued, for another, that blacks were more profitable: one could be maintained for four pounds a year as opposed to eighteen pounds for a white, and a black would work harder. Their availability for labor he thought would attract planters and tradesmen to the region. Nevertheless their servitude was not to be permanent. On arrival in America they were to be baptized and registered and equipped with dated dog tags. After seven years of working for others, they would be freed and given land at the rate of fifty acres per family in an all-black community in which whites would be forbidden to reside. He considered that blacks could become hardy, industrious, frugal, and loyal subjects.[94] This early experiment in apartheid was never attempted.

Phineas Lyman's tract of 1766 advocating a settlement on the Mississippi was similar in tone though more conventional on emancipation. It is a superficial question-begging work, in which Lyman argued that "the only method of settling such hot countries is by Negroes," on the slack ground that every European country with settlements in the West Indies had found it necessary to import labor from Africa.[95]

Bernard Romans discussed the issue in more depth. He had visited the Floridas in the early 1770s and returned to write a book about them in which he argued more cogently than Lyman that the experience of colonizing Georgia, from which slavery was initially excluded for seventeen years, had decisively shown that the "*primum mobile* of the welfare of these countries [i.e., colonies with hot climates] and of the wealth of their inhabitants are the African slaves." This contention was fortified, in his opinion, by the failure of the attempt in East Florida to work the land with free laborers from the Mediterranean, which Romans thought a "foolish" and "cruel" experiment. He warned prospective settlers in Florida that they could not hope to acquire wealth unless they became slaveowners. The "perverse nature" of blacks, he wrote, made necessary the harsh treatment they generally received. Stubbornness, idleness, and proclivity for theft were, he alleged, their natural characteristics and not the result of enslavement.

In interesting testimony to the impact that abolitionist literature was already having, Romans used several pages to attack *An Address to the Inhabitants of the British Settlements in America upon Slavekeeping,* an anonymous pamphlet then circulating in some of the colonies which combined the arguments of Montesquieu and of the religious abolitionists. In attempting to demolish the arguments of the pamphleteer, Romans refers to but does not directly address himself to Montesquieu, merely asserting that the Frenchman would not have argued against slavery if he had visited America and—pure assertion again—stating that the result of wholesale manumission would be a wholesale increase of vagabondage in society. Another unsupported opinion of Romans was that, working the same land in a hot climate, whites would produce ten times less than blacks. That Europeans could not work hard in heat was so well established, he wrote, as not to require elaboration. In opposing the Christian arguments against slavery, Romans cited several instances of slavery in the Old Testament and two in the New. Christianity, he was sure, did not forbid slavery but did require that slaves should be treated "as part of the reasonable creation."[96]

Among eighteenth-century writers on West Florida, Lyman and Romans were the only two who were prepared even to admit, however briefly, that there was an alternative to slavery. Both were northerners, and neither was a permanent resident of the southern colony. Lyman was an entrepreneur trying to organize other people for settling the province. Romans was a traveler. Another who journeyed into West Florida was William Bartram, who, as a Pennsylvania Quaker, might have been expected to condemn slavery there roundly. In fact he scarcely mentioned it in his book, and his few incidental references to it are not condemnatory. The complete absence of debate among West Floridians themselves makes it clear that they neither questioned slavery's existence nor their own practice of it.

Basic cultural, intellectual, and economic factors explain this acceptance of slavery. West Florida was in all senses a long way both from the fountain-head of "enlightened" ideas in western Europe and from the reforming notions which only some of the more zealous Christian sects had begun to espouse. The men of West Florida were not even in the main corpus of European movements, let alone of the vanguard. A new community, the colony of West Florida was traditional in thought and activity. There was little to stimulate intellectual diversity or vitality—no newspapers, no printing presses, no college, and only makeshift schools and churches, which might not have existed at all had the London government not provided stipends for schoolmasters and parsons.

There was minimal religious activity in the province. The little that existed involved Anglicans and, to a lesser extent, Catholics, neither of whose churches at the time questioned the institution of slavery. Although most of the colony's population was certainly illiterate, there were exceptions. Governor Johnstone admired Montesquieu, while the lawyers would certainly have read Blackstone. Even the semiliterate trader Fitzpatrick owned volumes by Raynal.[97] Books are owned without being read, however, and even reading an author does not ensure acceptance of his or her opinions. The callousness with which Fitzpatrick broke up slave families suggests that he did not accept Raynal's. A study of the colony in this period leaves the impression that economic survival and prosperity motivated the actions of people generally in West Florida. Johnstone's prime goal was to make his province flourish economically. To collect their fees seemed more important to officials and lawyers than almost anything. Traders and planters clearly saw that turning a profit was the purpose of their being in Florida, and the evidence, such as it is, suggests that all who were closely connected with West Florida thought the practice of slavery necessary for profit.

In such a climate one would not expect manumission to be at all common, and surviving colonial office records mention only a dozen instances in British West Florida. It is perhaps surprising to discover that William Gordon, in his very precise report on his parish of Mobile and the surrounding district, listed twenty-three freed blacks and mulattoes.[98] This suggests that there were perhaps four or five dozen freed men and women to be found in the entire underpopulated province. One at least was a property owner. She was Maria Belle, a freedwoman of Pensacola who, on 1 August 1773, while living in London, was given title to a town lot in Pensacola by Sir John Lindsay, a naval officer who had been stationed in West Florida in the previous decade when captain of the HMS *Tartar*.[99] Maria Belle paid no money, receiving the property only on condition that she improve and fence the lot, which was a good one, facing Cumberland Street and Pensacola harbor. A map of 1781 shows that it had a house on it, so perhaps she did carry out the stipulated improvement. Maria Belle was an exception. It seems that there were no other freed men or women of any substance in the colony, although there is evidence of two, John Paris and Stephen Williams, apprentices of Jeremiah Terry, who might have qualified as skilled artisans on expiration of their indentures. This qualification would not necessarily have set them apart from slaves who sometimes developed aptitude in cooperage or ship's carpentry.[100]

There was work for ship's carpenters in an infant industry for which West Florida was well fitted but which never achieved much size: in 1769 one square-rigged vessel of eighty tons was built there, in 1770 one ten-ton sloop, and in 1771 two small vessels which totaled twenty-four tons.[101] There would, however, have been a constant demand for ship's carpenters to repair storm- or worm-damaged vessels. Legally freed men and women had inferior rights to whites. Socially no doubt they were similarly accorded a low position. They were valued because in West Florida all population was valued, even slaves—perhaps especially slaves.

Importing slaves was generally encouraged as necessary to the West Floridian economy, but it was also seen as one of several ways to support the expenses of provincial government. For every slave over ten years old brought into the colony, the importer or purchaser had to pay ten shillings, and for every slave between fourteen and forty-five employed in the province, his or her master had to pay an annual poll tax of five shillings. To ensure that the authorities knew of all slaves brought into the colony, the master of any vessel carrying slaves to West Florida had, on pain of a hefty £100 fine, to make a full report of his cargo within forty-eight hours of arrival.

To encourage white immigration it was enacted that any slaves accompanying settler families or coming to the province to join settler households would be exempt from duty. So were slaves employed as sailors. Only if the slaves were subsequently traded was the duty payable.[102]

The returns for 1767 and 1768 have survived and show that the sums realized from duties were paltry.[103] In 1767, officials collected £10.10 from the slave import duty, and in 1768, £9.10, implying payment on only 21 and

19 slaves respectively. From the poll tax in the same years came £24.10 and £65.10, which suggests the existence of only 98 taxable slaves in the province in 1767 and a sudden rise in their number to 262 in 1768. This suggestion is surely false: we know that in those years West Florida contained far more slaves than even the larger of these two figures. But false too are the implications of the import duty numbers, if greater credence can be given to British customs office records than to the tax figures of provincial collectors.

The customs record indicates a decline in the number of slaves imported to West Florida between 1768 and 1773. For 1768 no precise figure exists. It is known that a schooner from Africa entered Mobile. One may presume that she brought in slaves, but the size of the vessel at only seventy tons precludes their being very numerous. In 1771 only twenty blacks were imported into Pensacola and in 1772 only nineteen. Neither of these two cargoes was dutiable, so the slaves were presumably not for trade.[104]

We know that the activities mentioned in these fragmentary customs records began earlier. For example, in a surviving extract from a letter written in January 1766 from Captain James Brimmer to a Mr. James South, the author wrote that he had three slaves to sell. Hector and Cato had cost him £60 each and Will £90, both in Jamaican currency. Brimmer wanted South to dispose of them in Pensacola or at any other port at which he might call for £100 each. If he was not overoptimistic, the captain would make a 30 percent profit on his venture for the expense of the short voyage from Jamaica to West Florida.[105]

Although generalization about slave prices in West Florida is dangerous, it does seem that Brimmer was unreasonable in expecting slaves to fetch £100 each in the mid-1760s. A price of $100 would have been more normal. Prices were governed chiefly by how much a slave could earn for his or her master, by scarcity, and by physical condition. Age, sex, and health were obviously relevant factors, as were availability and fluctuations in the West Floridian economy.

In 1766 the Mobile merchant Charles Strachan decided to auction his slave Will, provided he could obtain not £100 but $100. Evidently sickly, Will remained unsold.[106]

On another occasion a cargo of diseased blacks arrived off Georgia. The merchants of Savannah would have nothing to do with them, so they were sent on to Pensacola. The Floridians may have been short of slaves, but they were not so desperate as to accept men and women rotted through with yaws. The Pensacola merchants Jeremiah Terry and Evan Jones petitioned the provincial assembly to take action in the matter, with the result that a committee was set up which recommended that the consignment of blacks should be isolated on Santa Rosa Island or elsewhere to prevent their contaminating the healthy slaves in the colony.[107] Charles Strachan was supposed to sell three black girls from this cargo for a Georgia merchant, of whom one had died on voyage. Together the two survivors fetched only $183 and 8½ ryals.[108]

There was a good market for slaves on the Mississippi. By 1776, the last good year for British slave trading there, 900 slaves were brought to the river and sold for an average of £50 ($215), according to the perhaps exaggerated estimate of George Ross, a British merchant in New Orleans.[109] At the time, as we have seen, Bernard Romans recommended that potential immigrants from northern colonies take with them beyond their needs adolescent slaves for profitable resale on the Mississippi.[110] It was a plausible scheme only if they stayed healthy, because slaves, like everyone in the province, were prey to a variety of disabling or lethal maladies.

Yaws was very common. Originating supposedly in Guinea, it caused raspberrylike growths in the skin. When the disease was dormant the skin of an infected person would have spots which were perfectly smooth. Antoine Le Page du Pratz advised his readers never to buy a slave with the least hint of yaws, because it was generally thought incurable.[111] The other major diseases chronically afflicting blacks, according to Romans, were leprosy and elephantiasis, of which the prevalent variety in West Florida was characterized by grotesque swelling of the feet and legs. There was a common belief in the eighteenth century, and even in the nineteenth, that blacks were immune to yellow fever. Those who lived among them knew otherwise, and to guard plantation blacks against it, Romans advised owners to have them smoke tobacco and, before going to work, to drink a glass of herbal medicines soaked in rum.[112] Romans's solicitude implies kindness, as does the injunction of Du Pratz, "Take care of them when they are sick and give attention both to their remedies and their food which last ought then to be more nourishing than what they usually subsist upon," but it was probably like Dunbar's, strongly mixed with self-interest.[113]

One of the many disadvantages of being a slave was that masters who favored eccentric or even nauseating remedies for illness could impose their whims on their blacks. Opium and mercury were commonly prescribed, and Romans denounced the use of rum as "too often and indiscriminately applied to every disease as an universal *arca[n]um*."[114] Its effect, he wrote, was sometimes equivalent to poison.

A source of chronic trouble was that nobody knew anything about the origins of diseases. Remedies, however, were not always ineffective. Du Pratz's nostrum against scurvy, a mixture of herbs soaked in lemon juice, probably did some good in rectifying the vitamin C deficiency which causes the disease.[115] Slaves were also subject to smallpox, and in other colonies like South Carolina, owners were keener to practice variolation—an earlier and riskier form of immunization than vaccination, although a reasonably effective one—on their slaves than were whites to have themselves inoculated. There is no record of West Floridian slaves being inoculated, but the province was fortunate in being free from smallpox until 1770, when prompt measures to isolate sufferers saved the colony from an epidemic.[116]

Although the price of individual slaves varied considerably, especially because of physical condition, the slave prices listed in Appendix 4 were typical. Purchase by the batch rather than singly lowered the cost. Slaves with special skills were worth more than others. Children were considerably cheaper than adults, and mulattoes cost more than blacks. Indian slaves, incidentally, were uncommon in West Florida, but when available they were inexpensive.

Appendix 4, which lists ninety slave transactions occurring between 1764 and 1779—surely only a percentage of those which actually took place then—reveals much else besides. It demonstrates that borrowers very frequently offered slaves as collateral for repayment of loans. Certainly such debts did not necessarily imply financial failure, but having to surrender the collateral because a debt was never repaid probably does. There are several such instances which tend to show Pierre Rochon Junior and his brother Jean encountering severe financial embarrassment. Conversely, the lenders who collected the collateral were probably prospering financially, although not absolutely necessarily so, since it must have been sometimes much preferable to have money repaid than to have the trouble of caring for slaves. Such lenders included Alexander McIntosh and the partnerships of John Miller and Peter Swanson, of John McGillivray and William Struthers, and of Daniel and Benjamin Ward.

In three instances the list shows slaves manumitted either gratis or for a purely nominal fee; in two others Guillaume Loyson and Phillips Comyn freed their female slaves for $80 and $200 respectively, indicating that it was possible for slaves to accumulate during servitude more than nominal sums of money in West Florida. The apparent humanity of manumission is offset by the cold record that all of Robert Harley's female slaves were branded like cattle with his initials. Another seemingly callous man was René Roi, who gave the infant son of his slave Perrine to the freedwoman Charlotte, although knowledge of the attendant circumstances might excuse what is superficially shocking.

The table too shows that men employed in what eighteenth-century Britain regarded as humble capacities, carpenters and bakers, for example, were able to afford slaves in West Florida. The carpenter William Aird owned at least five, who were assessed at close to $1,500 and who learned their master's trade. Other slaves specialized in ship's carpentry and cooperage. At least three blacks worked for Guillaume Loyson and William Sanders, who, perhaps surprisingly, made their living as silversmiths in West Florida. Their slaves were not necessarily employed to learn their masters' craft, although a black silversmith is known to have practiced in New Orleans at the time.[117]

It is perhaps a little surprising, after the outbreak of war, when importing slaves by ship became intolerably risky, that there was not a significant increase in the value of slaves, but the detailed appraisal of the estate of Robert Farmar in 1780 at the end of the British period and of five years of war reveals no evidence of inflation in slave prices. York, who had skill as a cowboy, was appraised at $500, but ten adult males with nothing especially to commend them were assessed at $300 each. Most of the thirty blacks that Captain William Johnstone leased to the ordnance board in March 1781 were assessed at precisely the same figure: an exception, poor Blind Billy, was rated at only $200.[118] The standard assessment for Farmar's women slaves was $250 each.

At the time of his death, Farmar possessed fifty-seven slaves, but he and Johnstone were not the only large slaveholders in the province.[119] On their plantation at Baton Rouge, Stephen Watts and Samuel Flowers had sixty slaves. Farther upriver Richard Ellis had eighty-one on his.[120] William Clark, Augustine Rochon, and William Walker each had over thirty, as had, fitfully, William Dunbar.[121] Harry Alexander,[122] Philip Alston, David Holmes,[123] Harpin de Lagautrais,[124] John McGillivray,[125] William Marshall Senior,[126] William Marshall Junior, Alexander Murray,[127] William Murray, Elinor Neil,[128] John Newton,[129] Samuel Steer,[130] and Edmund Wegg[131] each had between twenty and thirty slaves.

Slavery in West Florida followed a pattern common to other southern colonies. As has been seen, Elias Durnford estimated slave numbers in 1774 at 1,200 as against 3,700 whites, or roughly one in three of the provincial population.[132] There followed the years of war, in which many whites immigrated. Often they brought slaves with them, but in many cases they were poor backcountry southerners who owned few or no slaves, and many too, although more prosperous, came from northern colonies and were not slaveholders. Whatever the case in the early 1770s, the years of revolution and war probably slowed the increase of the proportion of blacks to whites.

Contemporary opinion held generally that slavery was more profitable for employers than was indentured servitude, which did not thrive in West Florida as an institution, although there were perhaps a couple of hundred servants employed in various ways during its heyday. Romans assured intending

immigrants that buying slaves was the road to prosperity. They were thought
to be a better investment than land, even if, thanks to heavy settlement, it
doubled or tripled in value over ten to fifteen years.[133] All settlers were
probably aware of the comparatively recent experience of Georgia, where the
economy began to thrive only when slavery was permitted in 1750, eighteen
years after its foundation. That white men could not work as hard in the hot
southern sun and that blacks were less prone to sickness in southern colonies
were ill-founded prejudices, but they were omnipresent in West Florida. Since
abolitionist sentiment was unknown there and since there was a continuing
demand for slaves and, in time of peace, the means of supplying them, there
is little doubt that, had West Florida survived as a British colony, the proportion of blacks to whites there would have risen. To state that it was
impossible to make a living in West Florida without owning slaves would
clearly be false: there are too many examples to the contrary. To say that the
economy of West Florida would have developed much more slowly without
slaves is almost certainly true.

3

The Indian and Caribbean Trade

Like the phrase "slave trade," the words "Indian trade" evoke the past. Both types of commerce exploited human beings; though the Indians involved seldom complained of the trade's existence, they were often bitter about the way in which the white man conducted it. Long before colonization, from their very earliest contacts with Europeans, North American Indians bartered the products of field and forest and their own culture for commonplace European manufactures like nails and hatchets, items which were, for Indians, rare and inimitable treasures.[1] Over the centuries the Indians' self-sufficient way of life was eroded by a growing dependency on what Europeans supplied. By the eighteenth century Indians no longer hunted with the bow. Instead they relied on foreign traders to supply them with powder. Their families kept warm with blankets woven on machines in distant lands. They no longer dressed in skins but wore clothes sewn in Britain and Germany. Their cooking utensils were of iron that was dug and cast in the English midlands. The Indians' favorite drink was rum, which they had neither the knowledge nor the sugar to distill for themselves.

Barter remained the basis of the Indian trade during this evolution. In the eighteenth century Indians could obtain the European goods they needed by supplying deerskins and beaver pelts for the use of the clothiers, upholsterers, and hatters of lands across the sea.

The Treaty of Paris of 1763 improved British opportunities in the Indian trade. Previously the French had owned both shores of the Mississippi River, the natural route for skins and furs from the huge trans-Appalachian hunting grounds on their way to the sea. In prewar decades daring and aggressive English-speaking traders from South Carolina and Georgia had operated in what might legitimately have been claimed as a French preserve, but the diplomatic decisions of 1763 enhanced the potential for British traders by guaranteeing passage down the Mississippi to its mouth and by transferring to British rule the established fur-trading center of Mobile. Existing British trade could apparently now expand into a profitable monopoly among all the Indians of what are now the southeastern United States, including traditionally pro-French tribes.

As it turned out, several factors impeded the realization of that potential. One arose from a disparity between British and French practices in entertaining the Indians. The French had intermittently maintained good relations with the Choctaws who lived east of the Mississippi about 200 miles above New Orleans. At the French army posts in this region, the chief traders had been the garrison officers, who, fired by the prospect of gain, had entertained the Choctaws comparatively lavishly. By contrast, British army regulations forbade the officers who replaced the French from trading with the Indians, who proved sadly disappointed by the niggardly entertainment they received from redcoated officers, who had only their pay to sustain them. Adding to the Indians' discontent was the lavish disbursement, as late as 1763, of two years' arrears of presents by the departing French.[2]

For lack of an alternative the Indians had to do business with a segment of mankind less honorable than army officers—backwoods traders. They were the kind of rascally civilians described by James Adair, whose specialties included befuddling clients, using light weights, and profoundly despising all Indians. These backwoods traders were equally and chronically contemptuous of government regulation. It was feared, with justice, that, rather than sending the skins they acquired from the Indians to Charleston and Savannah for transshipment to Britain to the benefit of British colonial merchants, the economy of the mother country, and His Majesty's customs revenues, they would prefer to sell them to foreign entrepreneurs in Spanish New Orleans.

This traditional practice had only recently been made illegal. It was unfortunate for West Florida's economic prospects that a new zeal for stricter regulation of colonies afflicted the British government just as it was launching West Florida. The Plantation Act of 1764 (4 Geo. 3, c. 25), a wide-reaching statute that was broader than its popular name "Sugar Act" would suggest, hurt West Florida with additions to the "enumerated list." Products so listed might not be exported directly to a foreign market but could be sent only to Britain or to another British colony. Main exports of Florida, pot and pearl ash, timber, hides, and skins, all of which previously had been freely exportable to the French West Indies or to New Orleans, were, from 1764, on the enumerated list. The lowering of the duty on a beaver pelt imported into

Britain from sevenpence to a penny, which the act also specified, did not make up for the more general restriction.[3] Parliament included this reduction for the benefit of the declining English hat industry, not of Indian traders.[4]

A graver impediment to British expansion of the Indian trade was rivalry between the tribes. The Creek nation had warred intermittently with the Choctaws since the 1690s, were about to renew fighting in the 1760s, and had no wish to see their enemies supplied with British guns and ammunition. The Creek peoples were not united. They formed a loose confederacy with two main components—the Upper Creeks, who lived in villages along the Coosa and Tallapoosa rivers, and, farther east, the Lower Creeks, whose villages lined the banks of the Chattahoochee and Flint rivers.

The Creek chiefs of the early 1760s were astute. In 1763 rumors (later to prove accurate) that the British would replace the French in Mobile disturbed them; they knew that the change would give the Choctaws access to superior British trade goods, which formerly had to pass through Creek territory to get to the Mississippi country of the Choctaws. The Creeks had prized the control they had exerted over the flow of trade goods to their rivals and hated to lose such a good diplomatic weapon. In 1763, therefore, the chiefs put it about that the Creeks had never given Mobile to the French at the beginning of the century but had merely lent it to them.[5] As a tactic to persuade the British to relinquish the port, it was bound to fail. It served, rather, as justification for a powerful alliance of tribes to attack the British.

To resolve this difference and many others, the British government ordered a general congress for the southern tribes and representatives of the colonies bordering their lands. Organizing such an event posed problems. Convening numerous tribes at a specific place (in this case Augusta, Georgia) and time was difficult. Even getting them to agree to meet was not always easy. James Colbert, who carried invitations to the Creeks, found them in a particularly obnoxious and bellicose mood. They treated Colbert with "the utmost contempt and insolence," waving tomahawks over his head. A veteran trader of thirty years' experience wrote that he had never seen so dangerous a time.[6]

Particularly difficult to conciliate was the Upper Creek headman known as The Mortar, who had tried to organize a belligerent anti-British alliance with the Cherokees, some of whom had recently killed two British traders.[7] The Mortar swore positively that he would allow no trade goods whatever to go through Creek territory to either Choctaws or Chickasaws. Other Creeks thought that he was too extreme, and at Oakfuskie they took under their protection eighty packhorses loaded with goods belonging to the trading partnership of Brown and Struthers lest they fall into The Mortar's hands.[8]

In spite of all difficulties the Augusta Congress was a success. It began

properly in November with a sufficient number of representatives from all
the major tribes, the governors of Virginia, North Carolina, South Carolina,
and Georgia, and John Stuart, the superintendent of all southern Indians.
Stuart persuaded the assembled Indians that, contrary to mischievous rumors
spread by the departed French, the British wanted peace and intended no
punitive war against them.

Tribal spokesmen elaborated their complaints against British traders. First,
there were too many of them; second, because they talked young Indians
into exchanging deerskins for rum at an unfavorable rate, the young braves'
families lacked clothing. Third, traders did not confine themselves to the
villages but roamed the hunting grounds, disturbing the game.[9] Stuart replied
soothingly, but he knew that many of the complaints were valid and stemmed
from the government's policy of giving individual colonies both the right to
issue licenses to traders as well as the responsibility for disciplining them.
Giving licenses was easy and lucrative, which meant that colonial authorities
tended to grant them without considering the numbers issued by adjacent
colonies. Completely to control traders was all but impossible, as Stuart knew
better than anybody. He probably also knew that London was in the process
of devising a new policy to protect the Indians and supervise traders.

Although the Augusta Congress ended in apparent amity, some prominent
chiefs had refused to attend, and it was followed by more murders by Creeks.
Particularly shocking was the killing of fourteen whites in the Long Canes
region of Georgia in December 1763.[10] Trader arrogance that winter was
understandably replaced by fear. Some traders fled to the pro-British Chickasaws and some to the garrison towns of West Florida, fifteen of them finding
refuge in Mobile alone. By spring the consequent dearth of trade goods was
hurting the tribes,[11] demonstrating Governor Wright's truism that the English
could get along without the Creeks but the Creeks could not survive without
the English trade.[12] By April, promises of peace and good behavior lured
back the traders.[13]

The Upper Creeks, particularly The Mortar, were still set on damming
the supply of weapons to the Choctaws. That chief insisted that he wanted
no traders, unless they originated elsewhere than from West Florida, and that
he would regard any trade goods found in Creek country that had come via
Pensacola or Mobile as free gifts. He subsequently softened his position.
Perhaps because of the good effect made on Creek opinion by a couple of
young Britons from West Florida, John Hannay and Lieutenant Thomas
Campbell, who stayed in the Creek nation throughout the winter of 1764-
1765, The Mortar was persuaded to do what he had scorned to do before—
attend a congress.

This congress was held in Pensacola in May of 1765. There The Mortar argued with some logic that, since the Creeks had ceded lands to the English, the Creek hunting grounds were diminished and game had become more scarce. Therefore the price of trade goods in terms of deerskins should be reduced. At this point in his argument, the chief produced what he thought was a more reasonable price list.

In reply Governor George Johnstone denied that it was in the power of a single governor to lower prices. In any case, he continued, lower prices would deter the more respectable class of traders from commerce with the Creeks, who would have to rely on vagabonds and renegades.[14] His implication was that, where an honest profit was rendered impossible by low prices, only those who made their profit from swindling would bother to trade.

Superintendent Stuart then spoke. The traders were free men not slaves. He could not dictate prices to them. He could, however, ask his superiors in the British government for authority to make regulations for the Indian trade which would ensure just and fair treatment of both Indians and traders.[15]

The Creeks drew more comfort from his assurances than from Johnstone's words. They knew that, as far as they were concerned, his authority exceeded that of a single governor.

They were correct, but they did not know the complexity of the problems then facing Stuart.[16] By 1765 he was trying to implement a radical reorientation in Britain's policy toward the tribes which cut into established interests. This change was ordered in the celebrated royal proclamation of 7 October 1763, by which traders were to be made part of an imperial system. They would still be licensed by colonial governors, but they were banned from the forest; they might buy and sell only in designated villages. Also living in the villages would be imperial commissaries who could oversee their activities, and, if they broke the imperial rules, the traders stood to lose the security money they had deposited when issued with their licenses.[17] As a counterpoise to this strictness, the proclamation insisted that licenses were to be available to all applicants and without charge.

A step toward this system—which never was completely instituted—was a common code of conduct for traders in all colonies and a common price list specifying the prices in terms of skins and hides for the usual trading goods. Stuart was launching both code and price list in 1765.

The system outlined in the 1763 proclamation had been elaborated in Secretary Halifax's "Plan of 1764." Stuart liked it because it strengthened the independence and authority of the Indian departments but regretted that it did not permit him to license traders.[18] As a loyal servant of the crown, he of course tried to implement the policy.

At Mobile in March and April of 1765, he had scored a significant success. At a huge congress attended by Governor Johnstone, traders, 2,700 Choctaws, as well as Cusseta, Pascagoula, and Chickasaw Indians, he had secured general acceptance, after mature debate, of a twenty-one-point code.[19] Even as he hinted to the Indians at the subsequent Pensacola Congress, who included representatives of the Upper and Lower Creeks and smaller nations, that he would do what he could to lower the prices of trade goods, he must have known that it was all but impossible without forfeiting fragile trader support.

In the short run, therefore, prices remained unchanged. Consequently anger and frustration erupted among the Lower Creeks who met in November 1765 for another congress, this time at Picolata, near St. Augustine in East Florida. That the price of trade goods had not come down was the main complaint of those who attended. Other Indians said the same thing. "Rates too high," ran a message from The Young Lieutenant, a Creek chief who boycotted the congress; "If all the country were settled up to their [i.e., the Creeks'] towns, they would find nothing but rats and rabbits . . . deer are turning very scarce." Stuart, whose main purpose in summoning the congress was to extract a land concession, was disconcerted, but in the end, by skillfully using presents and medals as bait, he obtained the desired land for white colonization without the Creeks obtaining lower prices.[20]

Despite troubles caused by territorial exchanges, intertribal rivalry, and unprincipled traders, merchants exported their skins and furs almost from the first arrival of the British in West Florida. In this trade Mobile was initially more important than Pensacola, the capital, a dominance dating from the days of French and Spanish possession. The first known export cargo was a mere sixteen bundles, about 800 skins, sent to Charleston from Mobile in March 1764. A month later twenty-seven bundles of deer and one of beaver were placed aboard James Thompson's sloop *Adventure* in Pensacola. Although the export of 2,000 deerskins and 25 beaver pelts from Pensacola in July 1765 is worthy of mention, the first major cargo had to wait until August 1766, when 4,636 deerskins were sent from Mobile.[21] By 1772 deerskin exports were sizable. Official figures for the year, which excluded sales in New Orleans, show that 116,798½ pounds of dressed skins and 87,263 pounds of undressed skins were exported to Britain, while nearly 8,000 pounds of both went to other colonies.[22] Thomas Hutchins, a contemporary West Floridian, estimated that the annual shipload, in a 130-ton vessel from Mobile, was worth between £12,000 and £15,000.[23] The advent of revolution did not diminish the trade. In 1777 a London firm reported that it had two ships laden with furs worth a total of £40,000 in Pensacola harbor.[24] Despite the subsequent war with France and Spain and the consequent increased hazard of ocean travel, the fur trade did not die. The officer commanding the British troops in West Florida wrote to the plantations secretary that the merchants of Pensacola were prepared, if they were denied a convoy or escorts, to run the maritime gauntlet with their valuable cargoes of furs.[25]

All the cargoes just mentioned went from and to British ports. In the 1760s and 1770s a great quantity of skins were certainly also exported from West Florida to New Orleans, where the prices at times were higher and duties were nonexistent. Governor Johnstone estimated the consequent loss to the British treasury at £50,000, an incredible figure, although the total value of the skins traded was probably at least that much. It galled John Stuart that French traders, hangovers from the previous regime over whom he exercised no control, regularly traveled among the Indian nations, especially the Choctaw, to buy their skins, which normally they resold in New Orleans.[26] Johnstone's suggested remedies were for the government to remit all duties on the export of skins and instead to offer a bounty of sixpence apiece.[27] As with other expensive recommendations he had made to the government, these too were ignored, and the trade with foreigners on the Mississippi flourished.

General Thomas Gage, the overall commander of British troops in North America, was concerned at the blatant flouting of trade regulations, but no recourse was ever made to an obvious and often suggested check—siting a customshouse in that part of British Florida on the Mississippi. While many of the skins from up the Mississippi were sold in New Orleans and went to Europe in French or Spanish ships, exceptionally vessels brought Mississippi skins and furs to London. One was the *Florida Packet,* Robert Ross master, with 1,000 bundles of peltry aboard. Gage was sure that the snow must have been trading illicitly in foreign ports and caused inquiries to be made about her in London.[28] Two years later a cargo of furs worth £30,000 arrived in London from the Mississippi, part of whose cargo may well have passed through the hands of John Fitzpatrick.

Fitzpatrick was a middleman, one of many in a chain which stretched from the beaver building its dam on a tributary of the Mississippi to the fine felt tricorne on the head of a London dandy. His letters relating to the trade in furs and skins were written from 1768 to 1784.[29] Despite the assertion sometimes found that prices for deerskins were higher in New Orleans than could be obtained by exporting them to English markets, Fitzpatrick's regular customers for the skins he bought from Indian traders were British firms in Mobile, usually the partnership of John McGillivray and William Struthers but sometimes that of John Miller and William Swanson, who shipped them to England. That he traded with them from patriotism is most unlikely:

Fitzpatrick did not scruple to dispose of his skins in New Orleans when it was convenient to do so. He told the deputy superintendent of Indians in 1772 that in his heart he was a Spaniard by inclination and religion,[30] and he eventually became a Spanish subject.[31] That he traded with British firms because he was careless of profit is also out of the question. His letters over two decades reveal a consistent concern with maximizing gain. Selling skins in Mobile must then have been profitable.

Fitzpatrick did not like trading in skins. Several times he expressed the wish to give up that branch of business.[32] They were risky goods. If his skin consignments arrived too late to catch the annual boat that sailed to England from Mobile, the price went down. If he had to carry a stock of them through the summer, they had to be unpacked and regularly beaten to keep them free of worms. On occasion they were spoiled by water on the voyage from Fitzpatrick's trading house at Manchac to Mobile, while the insurance and freight payable on the complicated voyage added to the merchants' expenses.[33] Despite his distaste, Fitzpatrick had no choice. A mainstay of his business was to supply goods suitable for Indians to traders and hunters operating from as far north as the Illinois country. The shortage of cash in those remote regions meant that his customers could pay him only in furs and skins.

John Stuart believed that buying hides in hair enabled Indians to be cheated more easily than if dressed hides were traded, but in England was a demand for hides in the hair. It is no surprise to discover that the bulk of Fitzpatrick's stock was invariably of undressed hides. He also dealt in martin and beaver, getting a much higher price per skin, as much as seven reals (almost a dollar per pound) for beaver,[34] but his main business was in deerskins, of which the price varied in the decade from 1768 between thirty and forty-eight sols per pound.[35] The price for dressed hides was about five sols less per pound. The average weight of a deerskin was about two and a half pounds, but good ones weighed three.[36] They were normally bundled up in packs of fifty.

Sending cargo to Mobile involved two steps. Fitzpatrick shipped his skins downriver in the vessel of a "Captain Jerome," who was almost certainly Jerome Matulich, who plied the Mississippi. Somewhere along the river, either at New Orleans when the political situation allowed or, when it did not, at Rançon's plantation, near the mouth of the Mississippi, the cargo was transshipped to the vessel of Thomas McMin, who carried the goods to Mobile in the aptly named snow *Indian Trader.*[37] A simpler method of exporting skins than having to unload and reload them somewhere along the Mississippi was occasionally possible when what Fitzpatrick called "the New Englanders," probably the Nash family, would buy the skins from him, usually at a good price, to sell them in their native Rhode Island.[38] Exact totals are impossible to assess, but from the numbers of skins which Fitzpatrick mentions in his letters as stored in his house and, bearing in mind that he shipped them out as often as he could and certainly more than once a year, we can see that the volume of skins traded by this West Floridian was considerable. In January 1771 he had 8,000 pounds of skins in his house awaiting shipment. In May he had another 11,370 pounds. In January 1773 he had only 4,000 pounds of deer but expected a similar amount to arrive before the end of the month. The highest amount recorded is 16,000 pounds, which he had at the end of October 1774.[39] Thereafter, perhaps because of the disruptions caused by war, amounts are smaller—2,150 pounds in 1777 and 2,508 in 1784.[40] Fitzpatrick's letter book contains evidence of a continuing and, on the whole, thriving business in fur and skins, and there is no reason to doubt that the half dozen or so merchants of Manchac all participated in the trade in a similar way. They formed a lawless community, the province's wild west.

John Thomas, the deputy superintendent, indignantly reported that not one of them had a license and that they all sold rum freely to Indians, which was then against the law of the province. Fitzpatrick himself, according to Thomas, consistently swindled Indians, sometimes beat them, and had a penchant for dueling, not with the pistols or swords of gentlemen, but with the fusee, a light musket.[41] He and his fellows flouted the authority of Thomas with impunity. The deputy superintendent was not particularly well balanced, and in the end was recalled for himself dueling and killing George Harrison, one of the Manchac traders.[42]

The British government treasury papers available at Kew in Britain show, as surviving customs papers do not, how significant the importation of furs and skins into Britain had become in the years leading up to the American Revolution. Unfortunately they do not distinguish between the two Floridian provinces, although the amount from the eastern colony would be small compared to that from West Florida. Taken together, the Floridas exported, in the years 1773, 1774, and 1775, a total of skins (mostly deer) and furs worth just under £40,000. This amount was far short of what Canada exported and a little less than the export of neighboring Georgia, which was worth £47,000 in the same period, but more than the Carolinas and much more than Virginia, Maryland, New York, Pennsylvania, or New England.[43]

A breakdown of the types of furs and skins exported from Florida shows some interesting variations. Although the quantity in 1774 was much higher than in either the preceding or the succeeding year, and the amount for 1775 approximately twice that for 1773, yet there was a progressive increase for certain types of skin. While the amount of deerskins in the hair fluctuated, there was a consistent increase in the number of dressed deerskins and also, small as they were, in the number of elkskins, which rose from 26 in 1773 to 137 in 1774 to 141 in 1775. There was less consistency in the number of beaver pelts exported: 514 in 1773, 1,612 in 1774, and 17 in 1775. On a similar scale was the export of otter pelts, which numbered 267 in 1773, 539 in 1774, and 477 in 1775. Among the more exotic items listed were eight "tyger" skins in 1775 as well as twelve panther skins. (See Appendix 2.)

These Floridian export figures would be reasonably impressive even if they had included skins exported from the Mississippi region of West Florida. The figures in the treasury papers were compiled by the assistant inspector general of customs, John Tomkyns, and so we may assume that they were based on customs receipts. We may also assume that they did not contain estimates for the Mississippi not only because a customshouse on the river was lacking but also because the values assigned to the exports are not in round figures, as rough estimates must be, but they are precisely calculated down to the penny. The exports from the Mississippi would have been considerable. For instance, the 16,000 pounds of skins which we know Fitzpatrick stocked in 1774, which was perhaps less than his total sale for the year, were worth approximately £2,000 at English prices. The entire total of dressed and undressed deerskins imported from Britain, according to Tomkyns's incomplete figures in that bumper year 1774, was worth £19,603.10.0. So Fitzpatrick's stock alone, not to mention that of other Manchac traders, was worth just over 10 percent of what was officially imported into Britain from Florida in 1774.

Although he lived in West Florida himself, most of the skins Fitzpatrick dealt in originated elsewhere. One reason for the comparative paucity of West Floridian skins was that the rivalry between the Creeks and Choctaws had changed in 1763 into a war, which sputtered on with varying intensity until 1776, that is, for the bulk of the period of British rule.[44] Naturally the Indians killed fewer deer and beaver in wartime, and James Adair describes how his fellow traders often had to weigh the merits of the greater security offered to whites by timely war (as they saw it) against the greater volume of trade possible in time of tribal peace.[45] British government policy on this matter and others clearly affected the Indian trade.

In the later 1760s the ministry was looking for ways to reduce the expense of administering the colonies. The Stamp Act of 1765 and the Townshend duties of 1767 illustrate this endeavor. Less well known but with the same object was a change ordered in the management of Indian trade. In 1768 the new American department, on the recommendation of the board of trade, abandoned the system first ordered in the proclamation of 1763, which had made the trade an imperial responsibility run by imperial servants. Instead it would revert to the individual colonies for management. Backing its recommendation, the board asserted that the proclamation of October 1763 had never been intended as more than a temporary expedient—an admission made only because the result had not been satisfactory. Admitting the efficacy of Stuart's Indian policy in the south, the board noted that a similar policy had not worked well farther north. In spite of these arguments, economy was clearly a more important motive for changing policy. Between October 1764 and June 1765, West Florida alone had cost the Indian department the immense sum of 18,770 pounds and 19 shillings in New York currency,[46] roughly equivalent to £10,172 sterling, or, put another way, twice what it cost London to pay all the civilian officials in West Florida. From 1768 Stuart was forbidden to spend more than £3,000 sterling a year on Indian presents and other contingencies.[47] He and his northern counterpart, Sir William Johnson, kept their jobs, but there was no more money for commissary salaries.

Stuart was deeply disappointed. The authority conferred on him by the system initiated in 1763, despite its defects, had enabled him in 1767, much to Indian satisfaction, to persuade traders to lower their prices in exchange for a guarantee that he would grant trading licenses in general only to established traders and not to newcomers.[48] The predictable result of the new policy was to perpetuate ills which the 1763 scheme had been designed to cure, including in particular the proliferation of licenses to traders—lawless vagabonds is how the Creek chiefs at the 1769 conference described them—with badly defined and overlapping areas of operation.[49]

Stuart found himself tried. His responsibilities remained undiminished while the means to fulfill them were drastically cut. Not only were commissaries demobilized, but the redcoat garrisons, the symbols of imperial power on the frontier and the ultimate recourse for enforcing the king's will on whites and Indians alike, were now to be withdrawn from West Florida's interior forts. Again parsimony was a main reason for a policy change. It is doubtful, however, whether Stuart could have perfectly controlled the Indian trade, even if the troops had stayed and the commissaries he was establishing had remained untouched. His area of responsibility was huge and impossible to police effectively. If there were willing vendors and willing customers—and there always were, especially for rum—with a mutual interest in bypassing regulations, then illicit trading was bound to go on.

The extent of illegal trading varied. Rogue traders had to be more cautious when George Johnstone governed West Florida than when Montfort Browne was acting governor. Browne seemed totally uninterested in them, and in the latter stages of his tenure of office, no licenses were issued in West Florida, 80 percent of the skins brought to Mobile were traded for rum, and the traders themselves took the Indians' means of subsistence by themselves engaging in hunting.[50] Browne's temporary successor, Elias Durnford, was more responsible. The result was an act of the West Florida provincial assembly of 19 May 1770 which reintroduced a code of regulation together with a price list specifying the number of dressed deerskins required for any one of the commoner trade goods.

In this period, all codes compiled by southern colonial legislatures or governors for regulating traders and their Indian customers closely resembled Stuart's code of 1765, which he modified slightly in 1767 and which in 1768 he was prohibited from enforcing in all colonies, since it was not suitable for all of them. However ill-suited for the northern provinces, Stuart's code seems to have met the needs of southern colonies quite well, and several of them found it convenient to retain his code with minor variants to meet local needs. The West Florida code will be considered here.[51] Its local variant was a prohibition on selling skins in New Orleans or to French infiltrators.

To obtain a license, a trader had to put up as security for good behavior the hefty sum of £200 sterling. This license specified in which villages alone he might do business. On its back the trader had to certify all subordinates that he intended to use as clerks, factors, or packhorsemen. A trader had to agree to attend any congress convened by the governor of West Florida or the superintendent of the southern Indians. He was expected to supply intelligence that might be useful to British authority. He might not abuse or maltreat Indians. He could have a maximum of eighteen gallons of rum when in Indian country but could never sell any to Indians, neither could a trader sell rifles or swanshot, which spotted deerskins. Limits were placed on the credit which might be extended to an Indian, and he was discouraged from accepting undressed hides.

Enforcement of this act was predictably difficult. When the bond that had to be posted to obtain a license was high, the temptation not to get one at all must also have been high. The absolute prohibition on the sale of rum to Indians was quite unrealistic and must also have lessened respect for the act. The statute expired in 1772 and was not renewed, although Governor Chester believed that, without achieving anything like complete success, it had done some good.[52]

The coming of revolution shortly thereafter destroyed traditional trade arrangements. It became impossible to send furs from loyal Florida to rebellious Charleston and Savannah. For a while New Orleans remained as an outlet for British skins, despite the fitful severity of Spanish governors, but there too, from 1779, war ruined the trade.

In 1776 William Stiell, a half colonel of the Royal Americans, who was concerned with security and order in the Creek country, wrote to Governor Chester, pointing out that South Carolina, Georgia, and East Florida had all adopted a common code for the Indian trade and suggesting that West Florida should do likewise. It was the familiar modified Stuart code, though rather tougher on blacks and mulattoes. The price list that accompanied it is of interest as a guide to the Indian's life-style and priorities.

The most expensive item was a smoothbore trade musket which cost sixteen pounds of dressed leather. Equipping it with a score of flints, three-quarters of a pint of gunpowder, and fifty bullets cost another three. The next most expensive item was a duffel blanket which cost eight pounds of leather. Clothing was cheaper. A shirt cost three pounds of deerskin, a yard of striped flannel two, and a set of fifteen buttons one. A hoe cost three, a razor and vermilion to paint his face one each. A bridle for a horse cost three pounds of leather. The price of a saddle varied according to its quality.[53]

In most respects this list is identical with the tariff lists compiled by Stuart in 1765 and by the West Florida assembly in 1770. The main differences were minor decreases in the prices of gunflints, gunpowder, razors, and checked shirts. Although lower prices suggest a reduction of the Indian's cost of living, it should be remembered that eleven years of white encroachment on Indian hunting grounds had made it more difficult to acquire skins. Price stability probably also reflects the impact of more traders with more goods to dispose of than existed when they were all under Stuart's control.

In the long run perhaps the most significant aspect of West Florida's Indian trade involved not profits but the Indian's survival and the strengthening of the British empire. Indian trade proved to be the determining issue in the alignment that the Indians of West Florida took in the American Revolution. In the badly divided Creek nation, there was a strong neutralist element. The revolutionaries of Georgia and South Carolina tried very hard to make the neutralists prevail in tribal councils in the mid-1770s, but eventually the Creeks decided to support Britain because British traders could deliver goods more reliably than the Americans.[54]

War might have made the Indian trade more profitable for Britain if the renewal of war with Spain in 1779 had been followed by the British conquest of New Orleans, an often dreamed-of goal which local British forces lacked the means to accomplish. As it was, war hampered the development of the Indian trade.

The opposite was true of West Florida's trade with the West Indies. That close ties should have developed between them was natural. In travel time the province was closer to the islands, especially Jamaica, than to all the significant British colonies in the mainland. For defense West Florida was part of the responsibility of the British admiral at Jamaica, and with respect to communication the postal service provided a natural link. Packets from London sailed to Pensacola by way of Kingston, Jamaica, and then on to Charleston before returning to London. Most important of all was the interconnection of their economies. From the West Indies came the slaves which the labor-hungry mainland province lacked. In return West Florida could supply the lumber and provisions which the islanders needed.

Of course the West Indies had arrangements with New England for the supply of lumber and provisions, particularly fish, long before the Floridas became British, and strong commercial ties with the new provinces did not develop until the onset of the revolution broke the traditional connection. A certain brisk, if small and illegal, commerce certainly existed in 1771, when General Gage wrote to the earl of Hillsborough that some British ships seemed to have no other business than to trade between the Mississippi and the French islands.[55] As a small compensation to the devastation that it wrought on most of the American colonial trade, the outbreak of revolution elsewhere brought West Florida improved opportunities to trade legally with the British West Indies.

In 1775 Governor Chester wrote of trade prospects with a certain complacency to Lord Dartmouth, Hillsborough's successor as the plantations secretary. Already ships in the Mississippi were loading lumber and Indian corn. He would issue licenses to any persons from Jamaica to cut wood in West Florida. If the revolution continued the British islands could expect economic help only from West Florida and Canada. Since Canada would be isolated in winter, West Florida alone would enjoy a year-round monopoly of the resultant opportunities. The province lacked population—to increase which Chester offered several chimerical suggestions—but, he crowed, its lumber resources were inexhaustible. Within a year his colony could meet all the Indies' needs, and within ten West Florida would be able to supply tobacco, rice, and indigo too.[56] He wrote as though trade with the British Indies was a recent development, and if he referred only to Floridian exports, he was right.

Nevertheless there had existed dealings between the two areas ever since the British flag was raised in West Florida in 1763, for the initial British settlement was of soldiers transferred straight from Cuba.[57] The period of military government lasted just over a year. Civil government was established in West Florida in October 1764 by Governor George Johnstone, who came to Pensacola via Jamaica and, according to his own account, brought several Jamaicans with him. Certainly Johnstone had friends there because of his service in the Caribbean during the Seven Years' War, and he maintained contact with them. A grandiose scheme for importing camels offered him by a Jamaican acquaintance came to nothing, but less picturesquely, he did bring over plants and rum, while an important early visitor to West Florida was Sir William Burnaby, who commanded the Jamaica squadron and successfully took part in the crucial Pensacola Congress.

Chester was right in prophesying improved trade with the Caribbean, which was given a fillip on 27 February 1777, when the British House of Commons, sitting as a committee of the whole house, affirmed the expediency of granting a bounty on staves and lumber imported to the West Indies from loyal colonies, including West Florida.[58] Privateers, however, made the trade hazardous. On 3 May 1777 the sloop *Nancy,* under its commander, Robert Baker Gibbs, was brought into Charleston harbor as the prize of an American privateer, Captain Jacob Milligan's *Rutledge. Nancy* had been caught while carrying a cargo of lumber from Pensacola to Jamaica.[59]

Despite such depredations, the West Florida lumber trade continued to thrive. It probably reached its height in 1778. At that time a number of West Floridians, justifying a plea on economic grounds for more adequate defenses for their province, alleged that more than 100 vessels were employed in supplying the West Indies with lumber from their colony.[60] Self-interest may have caused them to exaggerate, but their main point was affirmed by a loyalist newspaper which commented in 1778 on the successful raid down the Mississippi of Captain James Willing of the United States Navy: "The Americans, with only 150 men, have totally destroyed the British settlements on the Mississippi, which are in every respect uncommonly promising, and from whence our West India islands have been principally supplied with the necessary article of lumber since the fatal contest with America."[61]

Among Willing's spoils was the brig *Neptune,* which was loaded with Floridian lumber destined for Jamaica. She was picked up in the Mississippi about thirty miles below New Orleans, and an early reaction of Governor Chester was to apply to the admiral commanding the Royal Navy squadron at Jamaica to send more vessels to West Florida for the defense of the great river.[62]

Willing also sent bateaux to the gulf coast to look for prizes. Although on the Mississippi these primitive vessels were normally propelled solely by oars, at least two of the three he sent on this duty were schooner rigged, meaning that they were equipped with a couple of masts and sails. One of them cut out the merchant brig *Chance* from her anchorage in Mobile Bay, complete with a cargo of staves, which the provincial government was unable to recover.[63]

As the tide of wartime fortune turned against Great Britain, her merchant vessels in the Gulf of Mexico became increasingly vulnerable. When Spain became her enemy in 1779, not only were the British settlements along the

Mississippi conquered, but the seas now contained the triple threat of American, French, and Spanish armed vessels. Undoubtedly the lumber trade was badly damaged, although it was not destroyed. Storeships sent from Britain had no choice but to chance running into enemy privateers and warships in the gulf. As late as February 1780 it was evidently their practice, after discharging their cargo, to load up with lumber for the return voyage, on which they would no doubt stop to sell the wood in the timber-starved Indies.[64]

West Florida's trade with the islands was by no means one-sided. In particular the Indies sent people to the mainland colony. The coming of war to the Bahamas early in the American Revolution caused several inhabitants of New Providence, as Nassau was then called, to look for greater security in West Florida. On 3 and 4 March 1776 the United States Continental Navy, in its first major operation, captured Fort Montagu, artillery, and Governor Montfort Browne, among other officials. Fearful of repeated raids, the Bahamian planter Thomas Hodgson successfully petitioned for land on the Pearl River in West Florida. The following month he sponsored a more general petition on behalf of his fellow islanders which the West Florida Council received favorably, although how many refugees took advantage of it remains unknown.[65] Many more emigrated from Jamaica. Often it was not free land for plantations they sought so much as the commercial opportunities offered by the Mississippi, which would give them access to markets in West Florida, Spanish Louisiana, and the British Illinois settlements.

The lack of customs returns from the Mississippi makes it impossible to assess precisely the extent of trade between the Caribbean and the settlements on the great river. It is perhaps significant, though, that, when Gálvez made his great raid on English vessels there on 17 April 1777, two of the eleven seized were from Jamaica. One was the brig *Hannah* and the other the ship *Peggy*. Another sloop seized then was the property of John Campbell, a Scottish merchant at New Orleans known to trade with the West Indies because, as previously noted, in the following year Willing took his brig, the Jamaica-bound *Neptune*.[66] One of the passengers made prisoner on that occasion was John Priest, a Jamaican whose reputation with the Spanish and British authorities (who called him a "dangerous bandit") was not greatly higher than Willing's. Like many another islander, economic distress impelled Priest to seek a better life on the mainland. He brought numerous slaves with him, whom he intended to sell in Louisiana. The Spanish government, however, suspecting that the blacks brought rebellious ideas with them, stopped him. Priest found buyers after he established himself on the British side of the Mississippi, at Manchac, with the cooperation of John Fitzpatrick, who was already there. Nevertheless, Priest's petition for 900 acres of land was refused, and Edmund Rush Wegg, the colony's attorney general, was instructed to remove him.[67]

In spite of privateers swarming in the Caribbean, the late 1770s saw a great many Jamaican firms trading with West Florida. In 1776 they included James and John Garnett of Kingston, Archibald Campbell of Hanover Parish, and George Woodhouse and the firm of Joseph Hardy and Thomas Ball, all of Kingston. In 1777 the Kingston partnership of David Duncomb and William Saunders traded with Pensacola. In 1778 we find the firm of Andrew Cathcart of Kingston and George Mowat of Savannah la Mar trading with Floridians, as did the triumvirate of Allen and John McLean and John Moore. Prospects diminished in 1779 after the Spanish conquest of the Lower Mississippi, but still Lewbridge Bright and David Duncomb continued to do business.

Trade did not always prosper and not just because of war with Spain. The sloop *Robert,* belonging to the firm of Thompson and Campbell, sailed from Montego Bay, Jamaica, for the Mississippi in 1777 but wound up a wreck on the Texan coast. There were only three survivors.[68] On 6 February 1777 it was reported from Charleston that the American brigantine of war *Defence* had just returned from a most successful cruise off Jamaica. Among five prizes taken by her commander, Thomas Pickering, was a sloop under Captain Munro which had been taken off the Grand Caymans south of Cuba while on her way from the Mississippi to Jamaica with a cargo of lumber. Later in the year, as previously noted, Jacob Milligan's privateer *Rutledge* took the Jamaica-bound *Nancy.*[69]

British ships also disrupted trade. On 7 February 1777, *Renown,* captained by Edward Hammond and owned by George Proffit of West Florida, sailed from Jamaica with permission to go to Guarico, where Hammond obtained a cargo for the Mississippi. (Guarico was in the French half of the island of Hispaniola—Saint Domingue, or, as it is now called, Haiti.) Arriving at the mouth of the river, the *Renown* was stopped and was made a prize of the Royal Navy vessel *Diligence.* Her commander, Thomas Davis, was evidently suspicious of the *Renown* because of some French crewmen taken on at Guarico. The *Renown* and its cargo were damaged and so much delayed that Hammond was unable to fulfill the contract made at Guarico. He protested the action of Davis to a court at New Orleans, with unknown results.[70]

Several Jamaicans of substance understandably chose to live in West Florida during the revolution. They could calculate that it was unlikely to become a battle theater unless Spain should join the conflict, and in that event Jamaica would be at least as vulnerable to Spanish invasion.

William Garratt Prescott was given land on Boyd's Creek on the condition that he bring his "considerable number" of blacks over from Jamaica. Less well received was John Laugher, who superficially (if his declarations were believed) was extremely loyal. He had lived from 1768 until June 1774 in Philadelphia, which he left because, he complained, allegiance and loyalty were looked on as criminal there. He fled initially to Jamaica but then to the Mississippi with a number of slaves. He was prepared to bring over some white and black servants. Laugher's potential contribution to West Florida may have been considerable, but the council rejected his request for 3,000 acres out of hand. Later in the same year, 1776, Alexander Maxwell was likewise rebuffed when he promised, if granted land, to return to Jamaica and to bring more than thirty slaves within three months.[71]

Even those who received land were usually unlucky in the long run. In March 1776 Thomas Dicas of Kingston bought 500 acres on the Mississippi between Manchac and Baton Rouge from the Mitchell brothers.[72] He ran up debts which he was preparing to settle by selling slaves when Willing's raiders burst down the river in 1778, carrying off and selling all his live property.[73] Dicas survived to become a member of the Manchac jury in 1780, but he still then had debts to John Davis and John Fitzpatrick dating from 1776.[74]

Willing similarly hurt Joseph Thompson, an even more recent arrival from Jamaica than Dicas, and the unfortunate firm of David Ross and Company, which lost a schooner and a valuable cargo to the rebels. Their vessel, the *Dispatch,* under its master, James McCrugh, had successfully voyaged from Kingston, Jamaica, to Pensacola and had entered the Mississippi by the southwest pass with a cargo of fifty picked slaves and 100 barrels of flour. McCrugh's intended destination was New Orleans, the headquarters of the Ross Company, but four miles beyond the pass, his vessel was seized by Joseph Calvert, one of Willing's lieutenants. Thereby Ross lost $21,450.[75] Equally unlucky was the Jamaican John Brown, who bought 750 acres on Houma Chita Creek in West Florida in February 1779, on the very eve of its conquest by Spanish invaders led by Bernardo de Gálvez.[76]

The possibility of slave rebellion in Jamaica, an island where blacks outnumbered whites by twenty-five to one, may have influenced some to leave.[77] Another factor may have been vulnerability to Bourbon attack, which prevented Jamaican resentment of the Stamp Act from developing into support for independence from the king and the protection of his navy.[78] Once the revolution was well under way, there was little doubt, to the relief of some on one side of the Atlantic and to the fear of some on the other, that the Bourbon powers would join the Americans. And if that occurred, there was even less doubt that the conquest of British Jamaica would be a war aim of high priority. Even if the dangerous and ruinous disaster of foreign conquest were averted—it ultimately was—the activities of the French and Spanish navies, added to the continued activities of American privateers, heightened the distress which the islanders were already experiencing.

There was no independent Royal Navy squadron patrolling the Gulf of Mexico. One had been suggested, but as the American Revolution turned into a world war in which Britain had no allies, she became desperately short of ships. The naval defense of West Florida depended on what the admiral at Port Royal could spare, which was never adequate. In addition the governor of West Florida, in his capacity as vice admiral, was empowered to issue letters of marque and reprisal to suitable vessels. Only one such is recorded. She was the *Orangefield,* a square-sterned, two-masted brigantine of eighty-three tons from Glasgow. She was armed with six carriage guns capable of firing a twelve-pound ball, two swivel guns, and two cohorns (a kind of mortar) which could hurl a four-pound shot at the enemy. Her owners were Glasgow merchants, and with a crew of thirteen she was bound to Jamaica when, at Pensacola in 1779, her commander, James Fairlie, asked to be allowed to operate her as a privateer against French, Spanish, and rebel shipping.[79] What success *Orangefield* enjoyed is unknown, but it is safe to assume that she was outnumbered by enemy privateers in the Caribbean.

Real hazards in Jamaica and the surrounding waters were to a large extent balanced by the physical dangers to be expected in Florida. It is true that a slave rebellion in the newer colony was a more remote possibility, although not unthinkable, as William Dunbar discovered.[80] But there were hostile and ill-disciplined Indians in the province, revolutionaries could and did launch raids, and the possibility of Bourbon invasion proved all too real. The Jamaican emigrants to West Florida were far from ill informed. As we have seen, they often made a reconnaissance before actually settling, and so they must have been aware of the dangers. That they came in spite of them indicates that they moved for economic reasons rather than asylum.

Trade between Jamaica and West Florida was evidently brisk during the revolution. One entrepreneur engaged in that risky business was Samuel Steer, who had been a considerable proprietor in Georgia until September 1775. When he rejected with disdain the offers of the Georgia revolutionary committee, life became difficult. On a schooner loaded with as much of his property as he could carry, he sailed for Jamaica, where he established himself as a merchant. For the next two years he traded between Kingston and the Mississippi, until in September 1777 he decided to settle on a plantation in West Florida with twenty slaves.[81] Steer did not abandon his mercantile ventures and ran into financial difficulties. As security for his appearance in the court of common pleas to face a suit brought by Philip Moore and

William Panton, he transferred a cargo to his guarantor, the merchant John Miller, which indicated the nature of his imports from Jamaica. These were contained in the schooner *Eleanor,* which had sailed from Kingston on 21 November 1778. The cargo comprised twenty barrels of regular sugar, one hogshead of long sugar, and twenty hogsheads of claret.[82] Another entrepreneur's run of luck did not last as long. The refugee shipwright John Davis arrived in West Florida in February 1778. Later the same year he set out from Mobile with a cargo for Jamaica. This schooner was taken by the revolutionary privateer *Dallas,* but Davis managed with some assistance—details are tantalizingly missing—to recapture the schooner and to take the *Dallas.* He sailed the privateer back to Mobile, but thereafter his schooner and its cargo were seen no more.[83]

Loyalists from mainland colonies fleeing from the revolution often arrived in West Florida penniless. Such was seldom the case for immigrants from Jamaica. Some were rich enough to become the creditors of established West Floridians. Cadwallader Morris, for example, arrived in Pensacola from Jamaica in 1776. On 13 April James Bruce, a prominent member of the West Floridian administration, acknowledged a debt to him of 806 Spanish dollars and 6 ryals. The security that Bruce and his wife Isabella offered Morris was two land tracts, one of 1,200 acres and another of 2,000, in the vicinity of the forks of the Amite and Iberville rivers, together with eight male and three female slaves and their children. Morris could hold them until 13 October 1777, the date on which repayment of the debt fell due. The collateral offered seems disproportionately large, but at least it gave Morris the opportunity to try his hand at planting in Florida without undue formality or permanent commitment.[84]

In a comparable if rather inferior position was John Stewart, who was variously described as "late of Jamaica" and a merchant of "Antego" (Antigua). Unlike Morris he had to borrow money but found kindly creditors—an amiable artillery captain, William Johnstone, and his wife Angélique. He was to discharge his debt of $3,000 by supplying the Johnstones within six months with three male slaves, two slave women fit for plantation business, and two twelve-year-old slave children. This agreement was much to Stewart's advantage in that it placed an unusually high value on the slaves. The Johnstones' generosity was again shown in their release to Stewart in September 1776 of 1,000 acres of their estate near Natchez. Stewart was allowed to offer the same land as collateral for his debt. Presumably the Johnstones found Stewart congenial and wanted to keep him as a neighbor. In the same month Stewart became owner of the snow *Peggy* of 150 tons. Stewart settled and did prosper, at least for a time. In April the following year he borrowed

$250 from John McGillivray. He offered nine black slaves as security and by the end of the following year had paid off his debt, with interest.[85]

Similar economic inspiration rather more blatantly drew several men of St. Vincent to West Florida at this time. The Windward Islands, without being technically neutral in the war, were thought of and referred to as neutral islands. The flight of settlers there who were looking to escape the tumult of revolution enormously increased the price of land in St. Vincent.[86] Harry Alexander, a middle-aged man of substance who had lived in St. Vincent for years, was a judge, a councillor, and the employer of twenty-five servants and slaves. Fearing that he could not provide land for his eight children on St. Vincent, he immigrated to West Florida. Clearly moved by economic considerations, he abandoned what was generally considered (quite wrongly, as it turned out) a refuge.

Other white immigrants accompanied Alexander. One was William Walker, also a councillor and a considerable proprietor. Leaving behind him in St. Vincent his wife, his three children, and thirty slaves, Walker arrived in West Florida with an overseer, a free black, and thirty-seven more slaves and was generously treated. In addition to 2,000 acres granted on family right, he received another 1,000 as loyalist bounty, all on a plantation east of Thompson's Creek, opposite Pointe Coupée. He was promised yet more acreage when the rest of his ménage arrived in the province. At least part of his family did follow him. In January 1778 his son Emanuel was granted 500 acres near the Great Lake at Tonica Bayou when he applied for land on behalf of himself and one slave.[87] Meanwhile the senior Walker had established himself as a trader on the Mississippi. Headquartered at New Orleans, he supplied Manchac merchants with Indies products like coffee and sugar.[88] The Spanish authorities favored him, and when the revolution, in which he suffered great loss, was not quite over, the governor of Louisiana allowed him to come from Jamaica to retrieve what he could from the remains of his business at New Orleans.[89]

A third passenger on the *General Morris* sloop, which arrived at Pensacola from St. Vincent in September 1776, was Thomas Colinson, whose family and slaves totaled nine persons. His reason for emigration was economic and more precisely described than most. A carpenter, the loss of lumber exports to the West Indies from New England had hurt him particularly severely. The captain of the *General Morris,* Levi Porter, also immigrated, bringing with him two indentured servants and six slaves, on the basis of whom he received 1,000 acres.

At the same time as Walker Senior's son applied for land, his son-in-law Thomas Hackshaw applied for 1,800 acres of adjoining property on the east side of the Great Lake near the Tonica Bayou. He left behind him on St. Vincent his wife, three children, and twenty-four slaves. He intended to bring them to Florida as soon as he could but hesitated while rebel privateers haunted the coast. The council was more interested in actual than promised immigrants and granted Hackshaw only 500 of the 1,800 acres he asked for.[90]

Another well-heeled immigrant was Thomas Durham, who came from Dominica. On 1 October 1776 he bought 500 acres of land on the Mississippi from Timothy Hierlihy for $250,[91] and later the same month he lent William Dutton $960, taking as useful security for repayment four slaves, Bob, Cato, Lucy, and Sylvia, as well as Dutton's 800-acre plantation near the Mississippi.[92] On January 1 of the following year, two of Durham's fellow islanders, Samuel Duer and Joseph Senhouse, bought three tracts totaling 2,100 acres on or near the Mississippi and Comite rivers for $5,600, a price which, considering the place and time, was very high indeed.[93]

Also from the Windward Islands was John Milner. If his own account is to be believed, he was bedeviled by misfortune. He had emigrated to West Florida in the very early days of British occupation, perhaps in 1763. He found life under military rule uncongenial and later went to England. He did not abandon his intention of settling in West Florida however, and finally in 1775 embarked with a cargo of slaves, plantation stores, and other goods totaling in value £5,000. The vessel containing them was wrecked in the Bahama Strait. His agent had neglected to insure them, and so Milner retired to the Windward Islands to gather his resources for another attempt at immigration. He found a suitable vessel, obtained plantation stores, and awaited only the arrival of a slave ship from the Guinea coast to supply him with slaves. News arrived that the slaves had been captured by revolutionary privateers and taken to Martinique. Therefore Milner sailed to Jamaica to buy the slaves he needed but found none suitable for his purpose. After leaving Jamaica on 12 May 1777, his vessel was captured by two provincial privateers off Cape Antonio. He lost all his possessions but was allowed personally to transfer to a Spanish vessel. Eventually he made his way to the Bahamas and finally from there arrived in West Florida. Perhaps his story was too lurid to be credible, for the council rejected outright Milner's request for 4,000 acres, but if his story were true, his suffering, whether from the weather, incompetent agents, or revolutionaries, richly merited recompense.[94]

There is little evidence of much trade between West Florida and Barbados, another of the Windward Islands. It is recorded, however, that early in 1777 the *Barbados Packet* of 180 tons under Captain Hawkins, carrying twelve carriage guns and twenty-five crewmen, was taken prize by an American armed brigantine while voyaging from Barbados to the Mississippi. It did not have a full cargo and presumably intended to pick up a load, perhaps of lumber, either in West Florida or Spanish Louisiana.[95]

Most southerly of the Windward Islands was Grenada, acquired by Britain in 1763. Robert Taitt left the island in 1777 because of the economic distress brought by the revolution. He arrived with nine blacks but stated his intention of buying several more. A slave dealer, he had suffered a severe loss when an American, Oliver Pollock, had captured a vessel containing 200 of his slaves. He was granted 550 acres on the northeast side of the Escambia,[96] where he was sufficiently respected to be elected to the West Florida assembly in 1779.[97] John Winnitt, an undoubted loyalist, came from New England. When General William Howe evacuated British military forces from Boston in March 1776, Winnitt left with him. He sailed for Grenada in October with a small cargo. He transferred his property, which by that time included five negroes, to West Florida in June 1778. In the twilight of the era of British rule at Pensacola, the council was comparatively generous. With much of the province in Spanish hands, it would have been unreasonable to designate a specific area. Not receiving a precise location for his land, Winnitt was granted 1,150 acres.[98]

Bermuda was tiny, 600 miles from South Carolina, and removed from revolutionary ideas and action, an unlikely source for immigrants to West Florida during the war. There is a record of only one such migrant, who arrived indirectly—the unusual case of Lieutenant Edward Jenkins. Following commissioned service throughout the Seven Years' War, Jenkins went to France for three years to learn viticulture, after which he chose to settle in Bermuda on the theory that, as its latitude and soil resembled those of Madeira, where vineyards flourished, it would be a suitable place for him. His winery failed, and he moved to Georgia, from which he was driven by the revolution. Initially taking refuge aboard the Royal Navy schooner *St. John,* he then migrated to West Florida. His bold request for 1,000 acres on which he would settle 100 whites and blacks was one to which the council did not care to respond. They postponed a decision. Later he lived in New Orleans.[99]

Very few French vessels ever visited Pensacola. In prewar years there was commerce between the French Caribbean islands and Louisiana, which for a time was illegal but nevertheless was practiced, much to the annoyance of the authorities at Pensacola, since French ships customarily flew British colors in order to enjoy unchallenged the navigation of the Mississippi.[100] Such vessels took pains to avoid the ports of West Florida. Occasionally, however, storms in the Gulf of Mexico compelled distressed French craft to put into British ports.

Such was the case of the brigantine *Languedoc,* commanded by Jean François Augias, which left Port Royal, Jamaica, on 30 April 1772. On 2 May she ran onto a bank fifty miles off the Jamaican coast, losing her rudder and part of her keel. Somehow she survived and on 30 May saw land. It proved to be Ship Island, west of Pensacola.[101] The *Languedoc* clearly needed major repair, and her cargo, with permission, was unloaded beneath the eyes of West Floridian customs officials. A thorough inspection revealed extensive damage. Augias doubted her seaworthiness and applied to the local court of admiralty for a judgment on the matter. If the court confirmed his conclusion, he requested that it should condemn her and thus enable him to dispose of her by sale in West Florida. A board of such ships' masters and carpenters as happened to be in Pensacola at the time surveyed the *Languedoc* and found her unfit for sea service: the cost of repair would be more than the price of a new vessel. And so the brigantine was condemned by court order.

Augias sought to pay the court costs and other debts incurred in Pensacola by selling part of his cargo, with the intention of shipping the rest to London aboard the *Planter,* commanded by Ephraim Wolfe. Such unorthodox export business could not be conducted without official approval, and the governor and council of West Florida considered the matter on 15 August. They agreed, with the stipulations that the cargo auction should be under the surveillance of customs officers and that, in accordance with British law, bond should be posted with respect to any cargo sent to Great Britain to make sure that the cargo, which comprised sugar and coffee, should reach its alleged destination.

James Jones, a member of the council who, with his brother Evan, had a business in Pensacola, begged for delay in transmitting this decision to Augias. He had relevant information which he wished to lay before the council at their next meeting on 17 August. Essentially Jones charged that Augias was a swindler. Two of his seamen, Jacques Grenier and Jean-Michel Furnier, swore in writing that, after *Languedoc* ran aground, Augias put into Jamaica instead of back to nearby Port-au-Prince in Saint Domingue so that he could covertly dispose of part of the cargo for his own private gain and charge fraudulent expenses to the *Languedoc*'s owner. Jones wanted all transactions delayed until instructions on the matter were received from France, a matter that would take months, at the very least.

Chester suspected that spite motivated this tactic: the Jones brothers were mortified at not being able to buy the cargo at a bargain price. He decided that Augias should be given a chance to answer the allegations against him. At the next council meeting on 20 August, Augias produced documented justification for all his activities, including the permission of the Jamaican governor to sell part of his cargo, a letter from several of his officers and seamen on the necessity for putting into Jamaica, and a certificate signed by the naval officer at Kingston of his necessary expenditures on repairs. He also had an affidavit from a passenger, Pascal Teste, who alleged that Evan Jones had tried to persuade him to use his influence to have Augias sell his cargo for 40,000 livres. Jones said he could not afford more, as he would have to pay people in Pensacola to obtain an export license—an interesting instance, if true, of the practice of "squeeze" at Pensacola.

Before overruling Jones, Chester took counsel with Attorney General Wegg and then gave Augias permission to do what he had originally requested. It is heartening to find that from equity Chester thus gratified a foreigner whom he would probably never see again at the expense of antagonizing prominent Pensacolans with whom he had constant dealings. The approval Chester received from the plantations secretary, however, must have been some recompense.[102] The only satisfaction in it for the Jones brothers lay in the evidently steep charges that they laid on Augias for landing, storing, and reloading his cargo. Augias's troubles, regrettably, were not over. The *Planter*, on which he sailed from Pensacola on 14 October 1772, was disabled crossing the Atlantic. She arrived at an Irish port in similar condition to the *Languedoc*'s on arrival at Pensacola. Another French vessel, the brigantine *Julie*, from Saint Domingue bound for Cape Fear, was driven by storm damage to take refuge in Pensacola in the spring of 1778.[103]

There was also a limited trade in lumber with French Hispaniola, practiced especially by former citizens of Louisiana who had migrated to West Florida. One such was Isaac Monsanto, who arrived to settle on the British side of the Mississippi in February 1770. Intending to export lumber to Hispaniola, he hired a snow of 180 tons. He contracted with three West Floridians named Parker, Bartholomew, and Richard Thompson to furnish 1,200 pieces of cedar, advancing them $500. Governor Chester proclaimed an end to cutting wood on royal land on 14 March 1771, whereupon Monsanto ordered woodcutting to stop, although 350 pieces had been already cut for export. John and James Durade, like Monsanto, immigrants from Louisiana and ambitious to export cedar, also asked if they could export what had already been cut for them by a Mr. Musto of Campbelltown. Both the Durade brothers and Monsanto received the required permission, on payment of attendant legal costs.[104]

The outbreak of war between the Bourbon powers and Britain did not end the clandestine trade with the French colonies, for much of it was conducted at a neutral meeting point, Guarico. On the other hand, the normal Jamaican trade with Europe became prey to French and Spanish interception in the later years of the revolution. It was facilitated by Jamaica's geographical location. Passage for merchant ships sailing through the narrow Windward Passage between Spanish Cuba and French Haiti was both awkward, involving much slow tacking, and hazardous. Because of prevailing winds, ships were compelled to sail westward around Cape Antonio at Cuba's western tip and through the Bahama Strait, separating the Bahamas from Florida, in order to reach the Atlantic. The Spanish maintained a lookout at Filipina, a high spot from which any vessels attempting to round Cape Antonio could be seen. A detachment of dragoons was always kept ready at Filipina so that news of such ships sailing could be sent to the Cuban capital, Havana, as quickly as possible. Although Havana was 120 miles away, readied naval forces could usually arrive in time to dispute a free passage to any English vessels heading for the Atlantic sighted from Filipina.[105]

There were few connections between West Florida and the Leeward Islands. Benjamin Graham, however, a merchant of Antigua, evidently had debtors in the mainland province. He empowered Richard Bradley, a Londoner who was going there, to recover his debts for him.[106] As usual, West Florida was a place more of debts than of dividends.

It seems that, like so many who forecast economic prospects, Governor Chester had been overoptimistic. The war did not bring a boom in Floridian trade with the Caribbean. The paltry quantity of cedar that Monsanto dealt in and the number of barrels of provisions imported by Samuel Steer and the types of vessels used—sloops and brigs rather than ships—are a guide to how small the scale of the trade was. All the same, there was a quickening of commerce and of immigration as a result of war. Had Britain managed to retain West Florida after the revolution, as she retained Canada, there is no doubt that both processes would have continued, particularly as the British authorities refused for some years after peace came in 1783 to allow the independent New England states to resume their traditional favored trade with the British West Indies. The Caribbean had much more to offer in that event than the Indian trade, since the ever-increasing pressure of white settlement into what had been British West Florida continued unabated after 1783, diminishing the hunting grounds in that region even further, destroying the game, and making it difficult even for the traditional hunters to survive.

4

Trading with the Spanish Empire

From its beginning as a British colony, far more profit was expected in West Florida from trade with mainland Spanish colonies than from either the Indian or the West Indies trade.

Flouting Spanish laws in order to exploit the profits of trading with the Spanish empire was a very old British tradition, practiced long before England herself had an overseas empire. In the sixteenth century, during his country's long war with Spain, Sir John Hawkins supplied slaves to Spanish settlers who were glad to buy them. James I's conclusion of peace with Spain in 1604 did nothing to abate Spanish opposition to English trade with the Spanish colonies. The Spanish did not accept peaceful coexistence "beyond the line"—the famous line of demarcation of 1494, which passed from pole to pole 370 leagues west of the Cape Verde Islands—to the west of which Spain claimed a monopoly on conquest, colonization, and trade. This grandiose claim was difficult to enforce, and British acquisition of Jamaica in 1655 enhanced opportunities to do business with Spanish settlers, who, hobbled by rigid trade regulations framed in Spain, usually wanted slaves and manufactures.

By the Anglo-Spanish treaty of 1670, the king of Spain recognized the existence of British colonies in the western hemisphere but made no concession to British traders, who continued commerce with his subjects even—one might say, especially—in time of war. During the War of the Spanish Succession, this trade intensified, and British triumph in the conflict was underscored, when peace was made at Utrecht in 1713, by His Catholic Majesty's assent to Britain's monopolizing the supply of slaves to his empire and, in addition, his permission for an annual ship full of manufactures from Britain to be sold at either Veracruz or Cartagena.

Despite these concessions, smuggling to the Spanish colonies persisted, contributing materially to the outbreak of the famous War of Jenkins' Ear in 1739, and continuing as usual during hostilities, flourished during the Seven Years' War, in the course of which a young naval officer, George Johnstone, saw much service in the West Indies and no doubt learned how mutually profitable a commerce unhampered by government regulations could be.

By the subsequent treaty of 1763, the British acquired Spanish Florida which they equipped promptly with garrisons but only slowly with civil governments. James Robertson, a British officer sent to reconnoiter the newly acquired territory, gained the impression that the Floridian ports would be declared "free and open to all nations."[1] In the short run he was correct, since initially there were no customs officials in the new Floridian colonies. The idea of free ports perhaps seemed good to Johnstone, who at the time was pursuing an ultimately successful campaign for the job of governor of West Florida, but he would be disappointed if he hoped that the peculiar status of his province would elicit tender consideration from the government in London. None of the various ministries presiding over the foundation and consolidation of West Florida as a crown colony in the 1760s proved sensitive to its special needs.

For every administration of the day, the salient aspect of any American colony was economic. The most detailed part of a new governor's instructions, including Johnstone's, concerned implementation of Britain's commercial rules for the province. Regulation had been piled on regulation for over a century, ever since the passage of the first navigation act in the mid-seventeenth century. That West Florida's geographical position gave it unique opportunities for trade with its Spanish neighbors and that its status as an infant colony might entitle it to laxer commercial restrictions than older colonies did not move the board of trade to modify its general policy, which, in the immediate aftermath of the Seven Years' War, emphasized enhancement of the British government's income through the more efficient collection of custom duties.

In 1763 Parliament passed "an act for the further improvement of His Majesty's revenues of customs and for the encouragement of officers making seizures and for the prevention of the clandestine running of goods into any part of His Majesty's dominions."[2] This statute was designed to make catching smugglers more profitable. It ordained that, after 1 May 1763, customs officials or Royal Navy crews who seized any vessel illegally importing goods into a colony were entitled to one-half of the profit resulting from the sale of the contraband; the other half would go to the British exchequer.

Throughout its existence British West Florida had a rudimentary and tiny customs service. It had no flotilla of revenue cutters to curb smuggling. By contrast the Royal Navy had many vessels in the area, and their officers, once peace was made in 1763 and they no longer could pursue prize money through the capture of enemy shipping, had ample time to enforce customs laws profitably. The squadron responsible for patrolling the Gulf of Mexico from the Mississippi River to Cape Florida was based in Port Royal, Jamaica. It comprised ten vessels, mostly sloops and frigates but including as flagship the fifty-gun *Dreadnought*. Their crews totaled over 1,500 men. In 1763 the vessels and the size of their crews were *Dreadnought* (a crew of 280), *Prince Edward* (220), *Venus* (190), *Adventure* (180), *Tartar* (160), *Active* (160), *Swift* (90), *Druid* (80), *Zephyr* (80), and *Lynx* (80), all of whom stood to benefit from catching smugglers.

The instructions of Rear Admiral Sir William Burnaby, who commanded Royal Navy vessels in Jamaica and the Leeward Islands, were that "no goods or merchandizes whatever" might be exported from or imported into any British colony, including of course the Floridas, except on a British vessel. To qualify as British a craft had to be built in Great Britain, Ireland, or the British colonies, it had to be British-owned, and at least three-quarters of her crew had to be of British nationality. Burnaby was required to use his "utmost endeavours" to enforce the parliamentary acts against smuggling.[3]

Since the natural trading partners of West Floridian merchants were the French habitants of Louisiana and the manufacture-hungry residents of the Spanish provinces bordering the Gulf of Mexico, strict enforcement of this hobbling restriction struck Governor Johnstone as ludicrous and contrary to the spirit of the British commercial system. All would have been well if British vessels had enjoyed free access to Spanish ports, but Spanish law was adamant in prohibiting entry to British traders. In practice the law was not always enforced, but there were enough ugly incidents involving Spanish coastguards and sensational accounts of imprisonment and even enslavement to make regular British trading visits to the ports of Mexico or the Spanish Indies out of the question, although, as will be seen, Spanish Louisiana proved more promising.

The instructions to the navy seemed beyond cavil. The wording used to prohibit importing and exporting in foreign vessels was directly lifted from one of the navigation acts, which for most Englishmen possessed the kind of sacred aura which the Monroe Doctrine had for Americans of later generations. In fact the instructions were disputable. The Navigation Act of 1660 had banned entry of foreign vessels into British harbors *unless* they carried bullion; the Navigation Act of 1696 repeated the prohibition without any exception. So which navigation act should be observed? Neither, thought

Johnstone, who wanted Spanish ships to have more access to West Florida than any existing statute allowed.

From the outset of his governorship, he tried to make Spanish purchasers legally welcome. His views filled the new colonists of West Florida with hope. On the day following his arrival, a Pensacolan wrote, "Our governor's presence will give a new turn to affairs: the step of sending away or seizing certain vessels with only cash on board is regarded as highly impolitick and which we are assured will not happen in the future."[4]

Three months later, Johnstone wrote at length to the secretary of the board of trade in London, enumerating the disadvantages of restricting the Spanish trade. He had, he alleged, received numerous letters from Cuba and the Spanish provinces of Merida and Campeche which indicated a huge demand for English manufactures, for luxury goods like furniture and saddles, but mostly for textiles—printed linens, checked materials, and baize. In exchange the Spaniards could offer dyewoods and bullion.

In the expectation that such a commerce would flourish, wrote Johnstone, "everybody" had crowded to West Florida when Britain had first acquired it. The numbers of immigrants in 1763, even before the arrival of a garrison to protect them, do support Johnstone's assertion that there was a rush, although some, like James Noble, were out for land rather than trade. But, continued the governor bitterly, over 200 disappointed traders had quit the province on discovering that the Spanish trade was "entirely interrupted by His Majesty's ships."

Another ill effect of the Royal Navy's zeal was the extreme scarcity of hard money in West Florida, so much so that bills of exchange had to be discounted as much as $23\frac{1}{3}$ percent, and even at that crippling rate, it was not always possible to obtain cash, "a thing incredible in a country surrounded with silver," wrote Johnstone.

Unless restrictions on the Spanish trade were abandoned, he continued, the capital would have to be moved from Pensacola to a site on the Mississippi at Manchac, where the land was rich enough to make the colonists self-sufficient and where the fur trade from upriver could be monopolized by the British instead, as was the case when he wrote, of falling into the hands of foreign merchants at New Orleans. The governor made it clear that, although such a move would be satisfactory in a limited sense, it would be a great deal more satisfactory if Pensacola could remain West Florida's capital. Not only was it suitably located for trade with the Spanish Gulf Coast provinces, but also, thanks to a waterway which Johnstone overoptimistically asserted was already open between Manchac and the gulf, it would garner all the fur trade from the Upper Mississippi.[5]

To bolster his plea for more leniency toward West Florida, Johnstone shrewdly enclosed two letters, both from senior men in professions requiring enforcement of Britain's mercantile system. One was from Commodore John Lindsay, then resident in West Florida, who argued that, unless restrictions on the Spanish trade were loosened, the new colony could not prosper in the near future. The other was from Hector Berringer Beaufaine, collector of customs at Charleston, South Carolina, to whom Johnstone left the tricky business of arguing that a literal interpretation of the navigation and trade laws would, if applied to West Florida, actually frustrate the goals of British commercial policy.[6]

Beaufaine began by citing the stated aim of the important Navigation Act of 1660, which was "encouraging and increasing the British shipping and navigation." That this aim was served by Spaniards bringing their tropical produce to British ports in America for transshipment to Europe in British ships he thought so far beyond doubt as to need no argument. He admitted that other words of the same act, that "no goods shall be imported into or exported out of" British ports in foreign vessels, did seem to imply a total ban but insisted that tradition favored a looser interpretation. Beaufaine cited the Sugar Act of 1733, which penalized the importation of foreign sugar into the colonies, but if brought in for the purpose of reexportation, it was not merely permitted but even made duty free. Similarly the Hat Act of 1732 had literally prohibited the export from the colonies of "any hats whatsoever upon any pretence whatsoever." Despite this wording, common sense, or "proper sense," as Beaufaine termed it, indicated that only hats made in the colonies were intended. Common sense too indicated that the freightage payable on Spanish plantation goods sent to Europe or on British manufactures sent in return, provided the transatlantic voyages were in British bottoms, promoted British policy. Such trade was with Great Britain, even if American ports served as entrepôts; "We are only factors," wrote Beaufaine. He scouted the possible objection that allowing foreign vessels to ship and land goods in British colonies would permit them to smuggle. That objection, he alleged, would imply that foreigners would have better smuggling opportunities or greater temptations than Britons, which he emphatically denied. In short, any chances for profitable smuggling were already being exploited by American colonists.[7]

The clamor against the 1763 act was widespread. A particularly pungent denunciation by a writer calling himself "The Englishman Deceived" appeared in the British press and was reproduced in American newspapers. "Can any Englishman be so mad as to stop a trade in which we receive, in return for everything we make, hard silver?" he asked. He concluded that the government had spent vast sums on a measure whose result was to compel the Spanish to fill French and Dutch pockets rather than British pockets with bullion.[8]

Some evidence supports Johnstone's allegations that West Floridians were losing money through the Royal Navy's zealous implementation of literally interpreted instructions. In June 1765 Midshipman Montague Blackwell testified that a Spanish vessel loaded with Campeche wood had been denied the right to land at Pensacola.[9] In London it was reported that during 1764 the exclusion of Spanish ships from West Florida had thrown £70,000 into the hands of French merchants at New Orleans.[10] On 18 June 1765 a Spanish sloop arrived from Veracruz at Pensacola, reportedly with $1½ million aboard. The Spaniards wanted to buy British manufactures, but no sooner had the sloop cast anchor than it was boarded by a naval officer with a squad of marines. Food and water were all that the Spanish could take on. If they tried to buy anything else, they were told, all the specie on board would be confiscated.[11]

Johnstone was not alone in protesting the exclusion of Spanish ships bearing bullion from British colonial ports. Even as he was marshaling the arguments for his pleading letter of 19 February 1765, in London the minister then responsible for the American colonies, Lord Halifax, was making a decision on the subject. In appearance it was uncompromising. Spanish vessels in distress or seeking food and water might enter colonial ports, provided that they "are not laden with or do not attempt in any manner to bring in any foreign goods or merchandize."[12]

This decision repeated the one made with respect to Jamaica on 6 June 1764 as a result of protests from the powerful lobby of the West Indies merchants.[13] In fact it was an ambiguous ruling, in that specie was unmentioned. Whether it qualified as goods or merchandise or neither was therefore left to the judgment of the local naval authority, who, as far as West Florida was concerned, was the admiral on the Jamaica station, Sir William Burnaby. Severer problems than the woes of West Floridians had preoccupied him on the Mosquito Coast in the early part of 1765, but on 10 May he arrived in Pensacola Bay with three ships of his squadron.[14]

The merchants of Pensacola waited on him, presenting him with a memorial which justly praised him for his work in the Bay of Honduras, but whose more pressing purpose was to seek his assistance in freeing the Spanish trade.[15]

Burnaby replied sympathetically. According to his understanding of both the letter and the spirit of the law, bullion qualified as neither goods nor merchandise. In addition he reminded the assembled merchants that Spanish ships could enter West Floridian ports for refreshment, provided that they were not laden with goods "to the detriment of British manufactures."

The final six words provided a significant modification of Halifax's ruling on 8 February. Since the logwood and dyewoods customarily traded posed no threat to British industry, the merchants were granted the substance of what they sought. Burnaby told them that he had already ordered the captains under his command of new rules to implement his interpretation of British policy.[16]

The position of the West Floridians was only temporarily satisfactory. If London ordered Burnaby to enforce its instructions more strictly or if the admiralty replaced him with a less sympathetic squadron commander, Spanish products would be excluded from Pensacola and Mobile; West Floridian merchants would then be like modern car dealers trying to sell new vehicles without being able to take used cars in part exchange. Business would decline sharply.

That the ministry needed to do even more to foster the Spanish trade was felt by manufacturers in England's northern and midland counties and most strongly by West Indian merchants, of whom several, like William Beckford and Rose and Stephen Fuller, were members of Parliament. The recent liberalization did not go far enough, they complained. Changes that then occurred were exclusively by executive order. They wanted a parliamentary statute which would place beyond dispute the right of all foreign vessels, not merely Spanish ships, to sell foreign produce freely and to buy British manufactures at designated ports in the British empire. That such free ports could enhance imperial prosperity was demonstrated, they alleged, by the Dutch boom port of St. Eustatius in the Lesser Antilles.

In 1765 John Huske, a Boston merchant, presented a paper advocating a line of similar free ports stretching from Pensacola and St. Augustine in the south to Portsmouth, New Hampshire, and Halifax in the north.[17] A coalition of West Indian and mainland parliamentary lobbyists finally pushed the Free Port Act through thinly attended houses in Parliament at the end of May 1766. The ports included four in Jamaica and two in Dominica. Pensacola and Mobile had as much right to inclusion as any West Indian ports, but no West Floridian port was mentioned, perhaps because the new law was hurriedly drafted and rushed through Parliament or, more likely, because the most active lobbyists had no Floridian interests.

Johnstone was deeply upset by the exclusion of Pensacola from the free port list; nature, as he wrote in a propaganda piece for the newspapers, seemed to have intended to create a center of commerce there. West Florida, he continued, perhaps sincerely, promised to be the emporium of the New World.[18]

Johnstone did not abandon hope of some concession concerning exports.

In a letter to the board of trade in October 1766, he asked that manufactures from continental Europe might be imported directly into West Florida without payment of duty if they were exclusively for reexport to the Spanish colonies (normally they could come to West Florida only by way of Great Britain). He specified textiles from Germany; lace, brocade, and silk from France; wine, brandy, and olive oil from the Iberian Peninsula; and from Italy, glassware, velvets, cards, paper, and "women's stockings made of coarse silk called capulto." Britain did not produce these goods, and if they were brought in British ships, the British economy would benefit. The alternative would be for the Spanish colonists to resort to English smugglers or to foreigners, in which case British authorities would gain nothing. Johnstone also sought permission to sell naval stores, pitch, turpentine, masts, yards, and bowsprits to the Spanish. Normally Britain alone was entitled to buy naval stores from her colonies. Even though the British treasury paid a bounty on them, Johnstone must have thought that they could be more profitably sold in Louisiana, where, if his request were ignored, he predicted that the Spanish would start producing their own naval stores. Finally, in order to secure for West Florida the furs of the Illinois country then being sold for high prices in New Orleans, the British government would have to remit duty on them and to offer a bounty.[19]

Nothing came of these suggestions. Even if the governor had not been out of favor with the ministry—indeed he was on the verge of recall—their extremity condemned them. It was unthinkable that West Florida would receive permission to supply a likely future enemy with the means of equipping its navy. In connection with European manufactures, his scheme proposed elimination of the profits of the mercantile middlemen who handled reexports from England; with respect to skins, Johnstone was asking the board of trade to decrease income at a time when augmenting it was a major preoccupation of the British government. Although his suggestions were ingenious, revealing unusually precise knowledge of what would sell in Spanish markets, none of them was apt to ignite support in any influential groups in England.

A few days before Johnstone wrote this plea, the effects of Admiral Burnaby's benevolence were seen when, according to the *Pennsylvania Gazette,* a large Spanish snow and an English schooner from the Bay of Campeche landed their cargoes at Pensacola without interference from the navy officers stationed there.[20] Their profitable cargo was logwood. The 1760s were a boom decade for the British textile industry. As a consequence there was a burgeoning demand for logwood, from which dyes, particularly the darker ones, were extracted. Also wanted in ever-growing quantity were tropical woods, mahogany in particular, for the craftsmen of fine furniture in London.

Perhaps at this time Jacob Blackwell, a customs official at Mobile, wrote an undated report about the potential profits from logwood. The trip to Campeche was, for the era, quick, and easy. Blackwell cited a schooner, probably the one referred to in the Philadelphia newspaper, which left Pensacola in July and was back in September with 119 tons of Campeche wood in company with a large Spanish snow carrying 200 tons of logwood which the owner of the schooner hoped to buy at £7.8.0 per ton. Since Campeche wood fetched between £10.10.0 and £17 a ton on the London market, a very reasonable profit could be expected, especially because, far from insisting on payment in cash, the Spanish wood vendors actually preferred payment in English manufactures or in West Floridian tar and pitch.[21] Blackwell's unknown correspondent may have been the Pensacola merchant Nicholas Talbot, who in 1766 rented the schooner *Expedition* for voyages to Campeche in order to load logwood.[22] In the following year contacts with Mexico fluctuated. On 25 January 1767, at anchor in Pensacola harbor were a Spanish snow, brigantine, and sloop.[23] Then for some months few Spaniards were seen there. The colonists believed they were absent because Louisiana's governor, Don Antonio de Ulloa, was relaying to his fellow governors at Veracruz, Campeche, and Havana the names of all Spanish captains known to visit West Florida. Ipso facto all were lawbreakers.[24] Repeated reports that there were not enough manufactures stocked at Pensacola to meet Spanish demand may have been a more important deterrent.[25]

In 1768 Ulloa was threatened with an internal revolt serious enough to secure his ultimate expulsion and had, therefore, more pressing concerns than reporting smugglers. Simultaneously the numbers of merchants in West Florida grew. In January it was reported that their storehouses were full, and in March that many Spanish vessels had called in at Pensacola.[26]

Montfort Browne, who acted as governor after Johnstone's departure, was anxious to encourage the Spanish trade, but in his greed for personal gain, he made the Spaniards wary of visiting West Florida. A Spanish vessel coming from Campeche to Pensacola was wrecked on the coast near the British port. Browne saved her cargo of ninety tons of logwood and shipped it to London, keeping the proceeds for himself. Using his unopposable influence in the vice-admiralty court of West Florida, he was also able to order the Spanish merchant to pay him for salvaging the ship.[27]

Thereafter, although a number of Spanish merchants continued to supply West Florida with indigo, cochineal, and logwood—all used in dyeing—in exchange for British manufactures, their numbers remained modest. There were three boats from Campeche and Veracruz at Pensacola on March 15, and the presence there of even a single Campeche schooner in June was newsworthy.[28]

There are no reports ever of shipping in Pensacola harbor from Cuba. Its governor was strict in enforcing Spanish trade regulations, even to the extent of firing on English ships trying to put into Havana for refreshment. In the first half of 1767 alone, there were at least three such incidents.[29]

Modest though it was, a trade conducted in Spanish ships definitely existed in the 1760s between West Florida and Spanish ports on the Gulf of Mexico. Undoubtedly beneficial to both colonists and mother country, it effected no modification of the board of trade's official policy. The royal instructions to Governor Chester, who was appointed in 1770, repeated without change the wording of the 1660 Navigation Act: "No goods or commodities whatsoever are to be imported into or exported out of any of our colonies or plantations in any other ships or vessels whatsoever but in such as do truly and without fraud belong only to our people of Great Britain."[30]

Chester had reason to resent the Spanish. When he voyaged to West Florida to take up his governorship in 1770, some of his personal effects went separately in the snow *Florida Packet*. Her master, Thomas Gallimore, had the misfortune to run his craft onto a rock off the coast of Spanish Hispaniola, where the colonists announced they were prepared to help salvage the *Florida Packet* only if given permission to loot the vessel. Chester lost his chariot and portraits of the royal family, among other possessions.[31]

On arrival in Pensacola he met David Hodge, one of the most active of the West Floridian merchants who traded with the Spanish. In 1770 Hodge visited Veracruz. His purpose was to help the colonists there export their bullion without having to pay the penal *indulto*, which literally meant a privilege but in fact was a heavy tax levied by their king when they exported bullion in ships of Spanish registry. Legally ships of no other nationality might be used. When Hodge offered a cheaper illegal alternative, he had the full approval of the British authorities, who appreciated the military intelligence which he culled at Veracruz.[32] Talented and adaptable, Hodge spoke Spanish like a native,[33] continuing to trade illicitly with Spaniards in Campeche for some years.[34] Eventually in 1785, after Spain had recovered possession of Florida, Hodge petitioned the Spanish governor for permission to continue living there.

Many others were to live and work under Spanish rule. Until its very end the eighteenth century was not marked by extreme nationalism, and Britons in the southern colonies were not deterred from seizing economic opportunities outside the protection of the union jack. Many chose to live in New Orleans under the French or Spanish rather than glean the disappointing fields of West Florida.

The paucity of records relating to British merchants in New Orleans during the French and Spanish periods of Louisiana history probably explains why the traders have received scant attention. Otherwise their neglect is surprising because there seems general agreement that they were extremely important. John G. Clark, the historian of the economy of New Orleans in the eighteenth century, has asserted roundly that from 1764 to 1768 "the English dominated the economic life of the colony."[35] In 1769 Governor O'Reilly wrote that 90 percent of the profits of Louisiana trade went to the British.[36] In 1776 Francisco Bouligny estimated that the British enjoyed an even higher percentage. More recently Gilbert Din judged that foreigners, of whom the British were the most significant, cornered three-quarters of the Louisiana trade.[37] The British merchants at New Orleans are worthy of attention, even if they are difficult to study. Probably because of war, expulsions, and fires, the letter books of only two of them have survived—those of Charles Strachan, who traded with New Orleans but who lived in Mobile, and John Fitzpatrick, who lived in the Spanish city for only one year. Fortunately they were a litigious group, and surviving court records throw additional light on their activities, as do governmental correspondence and records.

Many of them had a connection with the British colony of West Florida adjacent to Louisiana. It was natural for a British trader operating in New Orleans to maintain a second base in the neighboring colony, for he would receive, absolutely free but for legal fees, a grant of land in West Florida, provided that he was a British subject. He could double as a planter if he wished, and many did. If he preferred to confine himself to mercantile activities, a waterside lot at Manchac, Baton Rouge, or Natchez would be a distinct commercial asset. Even where there was no desire to own land in Florida, there was good reason to establish contact with the trading houses of Pensacola and Mobile because, in many respects, the economies of Louisiana and West Florida were complementary, while the British province had access to the variegated products of the British imperial system.

Whereas Merida, Campeche, and Veracruz had proved unamenable to British infiltration, New Orleans offered easy access. British vessels could voyage to it legally, either via Lakes Maurepas and Pontchartrain or, thanks to the free navigation clause in the peace treaty of 1763, up the Mississippi. Entrepreneurs sniffed profits. Even in northern colonies newspapers reported on New Orleans as a potential market.[38] Initially after 1763 the English visitors bought more than they sold, for until October 1764, West Florida was under a military governor who was ordered to send an expedition up the Mississippi to the Illinois country. It was outfitted at a cost of some thousands of pounds by two New Orleans firms, Logan, Terry, and Company and Isaac Monsanto, a future West Floridian.[39]

The activities of the military authorities in West Florida apparently con-
flicted with those of the British navy. The commander of the *David* sloop
stationed off Mobile was ordered to prevent trade between West Florida and
New Orleans.[40] So was Captain Lucas of the forty-gun warship *Princess,*
which was based at Pensacola. Together, or so it was reported in July 1764,
they seized three or four British vessels which had intended trips up the
Mississippi.[41] This anomalous naval activity ended only after the arrival of
Governor Johnstone.

Selling to the British was easy, but buying from them was not, because
in Louisiana, as its French governor Aubry lamented, with some exaggeration, "There is no longer any money, any commerce . . . the debtors no
longer pay."[42] Even as he wrote, Charles Strachan, the Scottish entrepreneur
from Mobile, was in New Orleans, where he sold $1,000 worth of textiles,
glassware, and gunpowder.[43] Presumably Strachan had adopted the trick,
ostensibly condemned by Governor Johnstone, of shipping his goods from
Mobile for Tangipahoa on the northern shore of Lake Pontchartrain—British
territory and thus duty free—and then covertly ferrying the cargo to New
Orleans via the Bayou St. John.[44] Nevertheless, even though trade with New
Orleans from West Florida was possible, it seemed clear that no fortunes
would be made while the penniless French governed Louisiana. Rumors in
1765 that the Spanish had decided finally to take it over occasioned large
consignments of English manufactures to Pensacola in readiness for an anticipated boom.

In spite of the changeover, dealings with New Orleans were at first minimal, for the Spanish evidently determined not to forbid trade but to control
it in a discouraging way. Governor Antonio de Ulloa, the first Spanish
governor of New Orleans, demanded that British vessels obtain a permit at
Balize at the mouth of the Mississippi before proceeding upriver. To obtain
it ships' masters had to produce invoices of all their cargo; the Spanish
authorities then set the prices at which it could be sold at New Orleans. It
was of no avail to avoid the Balize by using an alternative channel to reach
the Mississippi, for without a permit nothing might be sold in New Orleans.[45]

English traders were also deterred by the yellow fever and fluxes which
carried off many of the inhabitants of New Orleans in 1766.[46] Elsewhere,
though, the traders were active enough. Three convoys of goods from Philadelphia reached the Illinois country in 1766, to the detriment of French
merchants in New Orleans, who complained that the fur trade with the
Indians on which they had depended had fallen entirely into English hands.
It was some compensation for them that these Britons had a strong incentive
to sell their skins and furs illicitly in Louisiana rather than legally in Britain,

because furs then sold for 8 percent more in New Orleans than in London.[47] Nevertheless the New York press reported that all French traders were emigrating from New Orleans to Hispaniola.[48] This report was overoptimistic, but some time later, in March 1773, Israel Putnam did meet a schooner at the mouth of the Mississippi with forty French passengers who were leaving Louisiana for Cape François on Hispaniola because they were tired of Spanish rule.[49]

Growing internal resentment by the French inhabitants of New Orleans distracted the few Spanish officials from trade regulation. By the end of 1767, West Florida merchants were disposing at great advantage of large quantities of goods to the island and city of New Orleans.[50] They were also bringing lumber, bricks, and shingles from New Orleans in such quantities that, to discourage them, the West Florida legislature on 6 June 1767 passed an act imposing a hefty duty on the importation of foreign building materials.[51]

Even New York traders found it profitable to send cargoes to New Orleans during the period of transition from French to Spanish rule, particularly the years 1766 through 1768. A traveler who left New Orleans on 11 June 1768 reported that Captain John Walker's brig *York* was there then, while the brig *Africa,* which was owned and captained by William Moore, was toiling her way upriver toward the city. Both vessels had made repeated visits to New Orleans. Other vessels from New York at that period were the *Belle Savage* and the *Little Bob,* both mastered by Robert Harris, and Captain John Pell's brig *William.*[52] Not all New York adventurers were successful. Balthazar Kip lost his vessel *Kitty* on the Bahama Bank on 6 November 1767 en voyage to New Orleans.[53]

Flour from New York was in constant demand in New Orleans and fetched a high price. When the British first moved onto the gulf coast and some French moved out, a certain Jean-Baptiste Grevembert had sold Horn Island in 1764 to a consortium of British merchants for 625 quarters of flour.[54] Flour was the chief item of cargo of the *Africa,* the New York vessel which most frequently visited New Orleans in the late 1760s. Her owner, William Moore, first took her there in June 1766. Later in the year he visited New Orleans again and in 1767 made three voyages there. He filled large orders for the Spanish authorities, but they were slow to pay. Moore became impatient. Repeatedly he asked permission to go to Cuba to recover his debts in person from Ulloa's superior, Don Antonio Bucarelli. Moore received permission and finally sailed on 23 June 1768.[55] As passenger he took with him a messenger from Ulloa, Lieutenant Andrés de Balderrama, who prevailed on the Cuban governor to send some chests of coin to Ulloa in *La Hermosa*

Limeña. Even so, by 2 August 1768, a year after Moore had delivered the bulk of his flour to New Orleans, he had still been paid only one-third of the $25,000 he was owed.[56]

While waiting for satisfaction Moore of course had been compelled to pay his men and to maintain his vessel. Perhaps the combination of late and paltry payment overstrained his finances, for when next encountered, Moore was captain of a different vessel, and the *Africa* had changed hands. In 1769 Moore was captain but not owner of a brigantine of 140 tons, the *St. Peter,*[57] and by 1771 the *Africa* was the property of Thomas and Phillips Comyn, the first of whom was a London merchant.[58] Thereafter the *Africa* continued to frequent New Orleans but from the British capital rather than New York.

Another common visitor to New Orleans was Captain Thomas Hammond, whose sloop, the *Live Oak,* made voyages to the Mississippi from various American ports from 1764, including Savannah, New York, and Pensacola,[59] where, in 1768, he unloaded an unusual New Orleans export, mules.[60] They were probably destined for the Pensacola merchant John Stephenson, who ordered 120 mules originating from Natchitoches from his agent in New Orleans, John Fitzpatrick.[61] Hammond seems to have been on good terms with Ulloa, once carrying an important passenger for him to Havana, where the Cuban governor rewarded him.

Ulloa lacked the power to enforce Spain's commercial policy, although he could and did promulgate regulations. New trade rules for Louisiana were formulated in Spain in March of 1768. Nine ports of Old Spain were opened to imports from Louisiana. Duties on them would be a low 4 percent. From these nine, goods might be exported to Louisiana duty free, but only Spanish ships were allowed to carry them, and thenceforth no trade was allowed directly between Louisiana and any French ports. If this royal edict was designed to cure Louisiana's economic woes, it was a failure. There was almost no market for Louisianan products in Spain, while the European goods which Louisianans preferred, French wine for instance, were made in countries other than Spain, and importing them by way of Spain would make them inordinately expensive.

The regulations matched the needs of Spanish Louisiana as badly as did contemporary British regulations the needs of West Florida. The intended result of the 1764 Plantation Act, as has been noted, was to ban from West Florida's most promising markets, notably the foreign West Indies and Louisiana, all Florida's natural exports except indigo, which was already banned.[62] The difference between the provinces was that British merchants of West Florida had governors who neglected trade rules and encouraged smuggling, while the French merchants of New Orleans had a foreign governor who allowed smuggling only with reluctance and who was charged with publishing even more repugnant trade regulations than the British colonies had to endure. The Spanish edict was a dead letter. It turned simmering resentment into blatant defiance, and Ulloa was expelled from his province on 29 October 1768.[63] Retribution soon followed.

The arrival of a military governor, Allessandro O'Reilly, in August 1769 put a swift end both to rebellion in Louisiana and to tolerance for English trade. Unlike his predecessor, the general had the armed manpower to enforce Spanish trade regulations. He seemed shocked at the extent to which the English dominated the commerce of the province: "I can safely assert that they pocketed nine-tenths of the money spent here," he reported. He continued with obvious satisfaction, "I drove off all the English traders and the other individuals of that nation whom I found in this town, and I shall admit none of their vessels."[64] Since uncompromising exclusion was the official policy of Spain, it was natural that a governor should write in this vein to his superiors. O'Reilly perhaps was more sincere in his intended rigidity than other governors of Louisiana, all of whom had to live longer with the economic facts of life in the province.

In the end, according to the estimate of Bertram Korn, O'Reilly ejected only seventeen foreign merchants from New Orleans.[65] All the same, September 1769 was a very bad month for English trade with Louisiana because not only were English residents affected but also all the considerable number of English traders living elsewhere who did business in the Spanish town through proxies. One of these agents was John Fitzpatrick. A look at his letter book from the beginning of September until O'Reilly expelled him on September 22 is instructive.

Among other things the letter book shows what kinds of goods were then being supplied by English merchants at Mobile, Pensacola, the Illinois country, and Natchez to New Orleans. They included flour, slaves, strouds, blankets, checked cloth, and bear oil. It shows too what goods were in demand at the Illinois, from which Fitzpatrick's correspondent had floated down goods on a bateau. Fitzpatrick was asked to sell the vessel and remit the proceeds in tafia and barrels of pitch. Payment in kind was normal in New Orleans at the time, although in West Florida skins were more welcome than pitch, which was produced locally, but also acceptable were bills of exchange ultimately payable by respected British firms or the government. Fitzpatrick's letters also demonstrate that O'Reilly's exclusion edict was not as comprehensive as his official report would suggest. Any Britons married to Louisianans or who were planters were exempt.

Fitzpatrick had to wind up the affairs of eleven British firms or individuals in a matter of days. They included McGillivray and Struthers at Mobile; John Ritson, Valens Comyn, John Falconer, and James Amoss, all at Pensacola; Henry Le Fleur, Alexander McIntosh, Robert Barrow, and John Bradley at Natchez; and the Illinois branch of the Philadelphia firm of Baynton, Wharton, and Morgan.[66] Britons with whom he did not correspond at this time but whom Fitzpatrick mentions and with whom he had dealings are also numerous, and his letters do confirm the existence of many Britons with a commercial stake in the city. By 1769, however, the scale of their dealings was paltry. The largest outstanding debt mentioned was for $254. The figures would probably have been much larger if a desperate shortage of specie had not throttled a great deal of potential trade. The advent of O'Reilly, who presumably came with a treasure chest, was not therefore seen as an unmixed disaster by Fitzpatrick because, prior to the general's arrival, his Spanish debtors had been unable to meet their obligations and Fitzpatrick had lost potential customers because he was compelled to refuse Spanish paper.

Actually he saw commercial opportunity in the expulsion of all British merchants from New Orleans. There would be an increased demand for goods there, which Fitzpatrick intended to satisfy. He would operate from the West Floridian side of the Mississippi with a stock of $3,000 worth of textiles, mostly cottons. As close to the market as he was, Fitzpatrick must have felt confident of selling them. In exchange he was prepared to take almost all the items of local produce—deerskins, indigo, tobacco, rice, or raw cotton. He excepted lumber products as, presumably, too difficult to transport.[67]

Fitzpatrick was not alone in seeing the arrival of O'Reilly as reason for hope because of the half million dollars that he reputedly brought with him. The *South Carolina and American General Gazette* of 13 September 1769 described the general's coming as "good news for West Floridians." Another newspaper opined that it would be better for the West Floridians to have Spanish neighbors than French because "Great Britain has too often experienced the intriguing temper of the latter."[68] Lax Spanish enforcement of their commercial regulations was taken for granted, and many adventurers with merchandise of different kinds at once left Pensacola for New Orleans, hoping to exchange it for dollars. They were lucky because by 10 October, only a few days after the British merchants had been expelled, it was reported from Pensacola that O'Reilly had relaxed his initially rigid exclusion of British goods.[69]

Three months after his arrival, O'Reilly handed over the governorship of Louisiana to Luis de Unzaga, formerly colonel of the Regimiento Fijo de la Habaña. Unzaga was well aware that, though some trade with the English was desirable, the general effect of the large and growing British presence in the region was injurious to Louisiana's economy. Effective British control of the pine forests on the northern shore of Lake Pontchartrain barred Louisianans from the manufacture of pitch and tar. The amount of deerskins and fur locally available was vastly reduced because of the energy of British hunters. The Indians and habitants of the Spanish province also found it convenient to sell their furs to the ubiquitous British traders. Legal Louisianan exports were thus reduced to indigo, lumber, and in some years, rice, wrote the governor, and he railed against the "fraud and malversion" of the British traders.[70]

Unzaga's view of this subject may be contrasted with that of an Englishman who, as an army officer, had no stake himself in the trade. He was John Thomas, who served as deputy commissioner to the Indians at Fort Bute at Manchac on the Mississippi. Thomas admitted that there was much justice in the Spanish governor's complaints. Frenchmen from Louisiana who had bought Indian goods on credit at New Orleans were selling the pelts they acquired in exchange to the English merchants at Manchac and not returning to New Orleans to satisfy their creditors. But Natchitoches and Opelousa, both in Spanish Louisiana, were flourishing, in Thomas's opinion, because of the "vast quantity" of indigo and tobacco that Englishmen were buying there.[71]

Not only the Spanish in New Orleans had sometimes desperate needs which their compatriots could not but which the British could supply. In August 1770 two English brigantines with cargoes of flour arrived in the Mississippi. Unzaga persuaded their captains to go to Campeche on the Yucatan Peninsula, where the colonists were starving. Although Unzaga explained to his superiors that, if he had not diverted them, the flour would have been covertly sold elsewhere in Louisiana, he was reprimanded for his action.

It was a source of frustration to the governor that British vessels had a perfect legal right to sail under his very nose at New Orleans on the pretext that they were going to English Manchac and Natchez, when he knew full well that they intended to dispose of their cargoes to Spanish subjects. Usually he confined himself to warning them against illegal trade.[72]

Some of his warnings may have deterred merchants from planning voyages to the Mississippi, since the severe penalties for smuggling were well publicized in such places as New York and Rhode Island. Importers of cloth made from cotton or a cotton mixture in Louisiana, for instance, were informed that they were liable, not merely to have their cargoes confiscated, but also to pay a fine of twenty bits (that is, over two dollars) for every yard of illegal textile.[73] Despite all threats, British trade with Louisiana intensified, so much so that when Francis Murphy, acting for the Philadelphia firm of Barnard and Michael Gratz, arrived in New Orleans in the spring of 1772, he was disgusted to find such a glut of English goods there that, other than two saddles and some chintzes that he sold on credit, he had great difficulty in disposing of his merchandise. There were half a dozen vessels at anchor off the city when he got there, and more continued to arrive. Not only did Murphy have difficulty in selling, but he was also quite unable to buy any of the Louisiana products he sought. A French vessel had, shortly before his arrival, taken all the local produce available. He ruefully noted that the season for the sale of dry goods ran from August to March. In other words he had arrived to sell his goods precisely when the slack season had begun.[74] Murphy was not so discouraged that he never returned to New Orleans; in 1776 he had a residence in the town.[75]

Winking at breaches of the law for a time and then abruptly enforcing it was characteristic of English customs officials in Britain's North American colonies after 1767. Unzaga did the same in Louisiana.

One who suffered from this practice was the Rhode Island merchant John Nash, who, with his partner, Christopher Whipple, owned a sloop, *The Two Pollies,* which plied between their native colony and New Orleans with cargoes of New England goods. One such voyage began on 1 October 1773. The sloop was in the charge of her master, Ephraim Carpenter, probably one of a Newport family which did business with and settled in West Florida.[76] In the following month Nash and Whipple also left Rhode Island for the Mississippi in the sloop *Hope,* arriving safely in December. They found *The Two Pollies* anchored about six miles below New Orleans. After transferring *Hope's* cargo of merchandise to her, Nash then took *The Two Pollies* past the Spanish city and cast anchor a couple of leagues above it. Nash's story was the usual one that the goods were strictly for sale to English settlers in the various British settlements farther upriver. The Spanish doubted it. While at anchor on 17 February 1774, a Spanish sergeant and a corporal boarded *The Two Pollies* and asked to buy some codfish for their men in New Orleans. After the fish was weighed the sergeant begged successfully for the loan of a boat to carry it back to the town. With him Nash sent a sailor, William Proud, to bring the craft back. Instead of being allowed to do so, Proud was seized by Spanish soldiers and jailed. Nash himself then went to New Orleans to untangle the affair but was denied access to the governor. Instead of redress, on the morning of 20 February a squad of Spanish troops with drawn swords and fixed bayonets tried to board *The Two Pollies,* allegedly on governor's orders.

Being 3,000 feet from the shore, Nash thought that he could legally refuse but, in the face of force, allowed his sloop to be taken to New Orleans, where she was stripped of her rigging and her cargo, which, together with confiscated cash amounting to $5,760, were deposited in the royal warehouse. Pilfering soldiers and black laborers ensured that not all of the cargo got as far as the warehouse. The thieves seem to have been uncommonly literate and pious, for among the stolen items, if Nash is to be believed, were Young's *Night Thoughts,* Thomson's *Seasons,* Pope's *Essay on Man,* a *Book of Common Prayer,* a Bible, and French and English grammar books. Five days later Spanish officials appraised the cargo in the warehouse and *The Two Pollies.* On 4 March the cargo was sold for $482. Meanwhile Nash, Carpenter, his mate Benjamin Pilcher, and the crewman William Proud were all in jail.

On 2 May Nash was told that he was charged with illegally introducing fish for sale in New Orleans. He protested to Governor Unzaga. His petition was successful, perhaps because of a letter of remonstrance from Governor Chester of West Florida of 16 May 1774. On 17 June Unzaga revoked all proceedings against Nash and refunded $3,535 and 4½ ryals. He and his fellow prisoners were released after four and a half months in confinement; Nash and Carpenter had stayed not in the common jail but with a Mr. Murphy, perhaps the Francis Murphy who once sold saddles in the town. Together with the necessary expense of translating Spanish documents, Nash's lodging had cost him $385 and 5½ ryals. He estimated his total loss at $9,127 and 1½ ryals, of which he received scarcely a third in compensation. He asked in vain for a juster amount from Unzaga and then addressed his complaints to Governor Chester for transmission to Lord Dartmouth, the plantations secretary in Britain.[77]

This case illustrates several aspects of Unzaga: he was not vindictive, but he was negligent in supervising subordinates. Sergeant Hildago and his corporal were both dismissed from the Spanish service for their part in buying English fish. The case also suggests that there was probably a regular commerce between Spanish Louisiana and Rhode Island. Thanks to this case too we know in detail the kind of cargo which sold along the Mississippi, certainly to English settlers but also, probably, to habitants on the western shore. The vast bulk of the cargo was an amazing variety of textiles, much of them coarse and cheap, all of them from Europe, mostly from England. Such would certainly have been the origin of the beaver hats which were part of Nash's cargo, since he would scarcely have declared in a memorial to the British government items that could not legally be manufactured in the colonies. The sum of money confiscated from *The Two Pollies* was large enough to suggest that Nash had probably engaged in some trading with

Louisianans before her seizure.[78] The *Two Pollies* episode illustrates the misfortunes attending entanglement with the Spanish authorities. If an indignant press may be believed, it was not an isolated instance of Unzaga's rigor. Two days before, dry goods to the value of $12,000 belonging to a merchant named Basset had also been seized in New Orleans.[79] Undeterred by his treatment Nash remained on the Mississippi and by February 1778 was an inhabitant of Manchac.[80]

Other British traders were also persistent. Shipping lists for the year 1774 show that well over a dozen voyages took place to the Mississippi from mainland colonial ports alone.[81] The coming of revolution to the mainland colonies in 1775 did not end traffic with the Mississippi, but it certainly did introduce several new factors into the trade.

For example, ever since the foundation of West Florida, guns, ammunition, and powder had been exported to the Mississippi. For defense and hunting they were a necessity for white settlers and Indians alike. The imaginative George Johnstone had sought without success to stimulate local production of gunpowder.[82] When the quarrel between Britain and her colonies turned into war, George III found it necessary to restrict the export of munitions to North America. The result was that the provincial authorities began in 1775 to intervene in what previously had been considered purely private transactions. In November the ship *Ann*, William Reid master, arrived from London at Pensacola. Its cargo, 9,000 pounds of powder, 925 Indian guns, and a quantity of bullets, had been ordered by James Mather, an English merchant residing in New Orleans, and intended for sale up the Mississippi. The West Florida council ordered it unloaded and stored in the provincial magazine at Pensacola, after which 1,000 pounds of the powder might be sent to Manchac, not by way of the Gulf of Mexico, where it would be liable to interception, but via the lakes and the Iberville. Its sale on the Mississippi would be supervised by Thomas, the previously mentioned deputy Indian commissioner. After its use, another 1,000 pounds could be sent to the Mississippi.[83] The same rule was applied two months later when the brig *Norton* arrived from London with 2,000 pounds of powder and a similar quantity of ball. Captain William Pickles asked for but was denied permission to carry it directly to Manchac, a sensible precaution in the light of Pickles's later pro-American activities.

The Continental Congress was undoubtably keen to obtain military supplies from New Orleans by way of the Mississippi and Ohio rivers. In 1776 a large barge containing nineteen men and a boy left Fort Pitt on the Ohio, sailed into the Mississippi, and arrived on August 1 at Walnut Hills (Vicksburg) in the northwest corner of West Florida, where the American flag was raised by their captain, George Gibson. He carried dispatches from Congress to Governor Unzaga and to his royal master, but Gibson also wanted to trade his cargo (presumably of skins) for gunpowder. The success of this type of trade, about which Unzaga was only initially cautious, worried the British authorities.[84]

Competition for Spanish customers increased, to the detriment of British Mississippi traders in 1777, thanks to a Franco-Spanish agreement of the previous year which Bernardo de Gálvez, who succeeded Unzaga as governor of Louisiana, had to implement. Exports to France and the French West Indies were made legal on payment of a 5 percent export duty. Direct imports from France were also permitted, thus eliminating any further need for the French to resort to subterfuges like purchasing vessels of British registry to practice commerce with Louisiana.[85] Now slaves could be sent from the French West Indies in payment for the products of the Spanish colony. John Fitzpatrick of Manchac saw a gloomy future for British rivals to the French traders: "They certainly undersell us and their goods are better calculated for this province," he wrote.[86]

Nevertheless there was no hint that Gálvez would suddenly and vigorously stifle existing British trade with Louisiana. One writer noted that successive governors had "for years past," in return for a small share of the profits, connived at such activity.[87] The word on the river was that Gálvez was even more liberal toward English traders than his predecessors. "The new governor," wrote Fitzpatrick in February, "allows the English liberty to trade or hunt up any of the rivers on the Spanish side they please; further—all the English merchants that kept their stores on board the vessels have now their shops in town."[88] The sense of security into which they had been lulled must have been profound indeed if they had abandoned the sensible practice of maintaining their floating warehouses in favor of building in New Orleans, but disillusion was soon to follow, as Gálvez reversed his tolerant policy.

It is now half a century since John W. Caughey analyzed Gálvez's motives for this policy reversal. They included the new instructions opening his colony to French traders which simultaneously ordered the exclusion of the British, Gálvez's discovery that Louisianans were ready to inform against Britons, and his pique at a new British insistence on seizing small Spanish vessels violating British regulations on Lake Pontchartrain.[89]

Lieutenant George Burdon had been legally correct but perhaps overofficious in seizing, early in April 1777, two schooners going from Bayou St. John to the Pearl River at the eastern end of Lake Pontchartrain. They were smuggling wine and tobacco; one of them had 160 "sticks," or "carrots," of tobacco, or something less than three-quarters of a ton.[90] Had they not been caught, the schooners presumably would have loaded up with tar or staves for the return voyage. This legal justification appropriately prevailed in Gálvez's explanation to the governor of Cuba of his retaliation.[91]

Caughey might have added two more reasons for the policy change. One was economic, namely, the draining effect on the Louisianan economy of the one-sided trading with the British. Louisianans sold little but bought much from the Floridians. One estimate is that the habitants had an annual import bill of $700,000.[92] Exports, according to Francisco Bouligny, were smaller: $100,000 worth of lumber, $180,000 of indigo, and $200,000 of peltries; with the addition of such less important extras as mules and bear oil, a total of perhaps half a million dollars. Bouligny estimated that, apart from $15,000 of it going to the Spanish and a smaller amount to the French, the whole of this trade swelled British accounts.[93]

The other reason for the Spanish change was political—the effect of the revolution. New Orleans was inhabited by both loyalists and rebels and was increasingly disturbed by English-speaking rival partisans who brawled in the taverns. Spanish citizens complained that there was "no time of the night or day in which they were not alarmed with screams, blows and cuffs besides private challenges." Gálvez decided that he could not permit "this petty civil war" in the heart of his capital. One center of strife was the boardinghouse of Hannah Ogilby, who welcomed loyalists only. Her outspoken condemnation of rebel politics, according to her analysis, caused confiscation of her property and detention for twenty-four days in a Spanish jail; finally Hannah fled to Pensacola with her small daughter.[94]

Political considerations apart, the English residents of New Orleans were of low caliber, "vagabonds and bankrupts," wrote Gálvez, who offered as proof the fact that British prisoners in the New Orleans jail outnumbered Louisianans by two to one. All of the former were guilty of crimes against society. Transient Englishmen behaved no better than residents. From their vessels they ran amok among the plantations of Louisiana, taking off slaves, killing cattle, and shooting at planters trying to protect their property. After these forays they would retire to the sanctuary of their boats anchored in the neutral waters of the Mississippi.[95] Naturally Gálvez did not mention American influence on his decision, but he was not the first governor of New Orleans to favor American patriots.

In 1776, as already mentioned, Unzaga had shown partiality to George Gibson, who had sent a bateau up the Mississippi with a Spanish flag, a Spanish pass, and 8,600 pounds of gunpowder which eventually reached the Ohio for use by revolutionary troops in the spring of 1777. The bateau had wintered at a Spanish post on the Arkansas River. Gibson himself had sailed with nearly two tons of powder in a sloop under George Ord, again under Spanish colors, and safely bore it to Philadelphia, where it was received by the firm of Willing and Morris.[96] Oliver Pollock, the American agent at New Orleans, had supplied the powder. Unzaga, who sold it to him, justified disposal of these Spanish government stores as old and partly spoiled.[97] Gálvez's pro-Americanism was as pronounced as Unzaga's, understandably so, since their royal master repeatedly ordered him to favor the rebels.[98] In April 1777 a Spanish vessel at New Orleans was allowed to fly at her topgallant masthead and *flanked* by French and Spanish colors "a flag in which was a snake and a hand grasping thirteen arrows and the field divided into thirteen stripes of different colours."[99] The corollary to this partiality was hostility to Britons. When Gálvez finally decided on action against them, he was drastic and swift. On 17 April he ordered the simultaneous seizure of all British vessels between Balize and Manchac and, on the next day, commanded all English merchants to quit his province within fifteen days.

At the time there were thirteen vessels on the river. Two escaped. The Glasgow brig *Jesse* was then under sail and could not be stopped. Another vessel bound for Britain was not moored accessibly. She too got away, so the Spanish boarding parties took eleven craft with English-speaking crews. The masters of the two most valuable, Joseph Calvert and William Pickles, of the brigs *Steady Friend* and *Norton,* turned out to be Americans. No doubt Gálvez would have left them alone, but distinctions were hard to make when the Americans carried English passports and flew English colors.[100] Pickles alleged that he had Unzaga's permission to trade with Louisiana. Subsequently both he and Calvert were given back their vessels.[101] Of the larger craft permanently seized, the brig *Hannah,* owned by Archibald Dalziel of Jamaica, was in ballast and on her way upriver. Another Jamaican, Captain Collart, owned the sloop *Peggy.* The remaining seven vessels—three brigs, two sloops, and two schooners—were owned by British Mississippi traders well known in New Orleans. Of these the major sufferer was the partnership of George and Robert Ross, which lost the brigs *Hercules* and *Camilla.* John Waugh, the seafarer from London who had made the Mississippi the theater of his commercial operations, lost a brig and a sloop, both full of cargo. John Campbell too lost a sloop. The firm of Patrick Morgan and James Mather lost an ancient schooner, together with a comparatively valuable cargo of twenty-two slaves.[102] A similarly wretched vessel was the schooner *Sally,* owned by David Ross.

The Spanish authorities sold all their captures at auction or, as the current phrase went, at "vendue." The vessels themselves realized a total of $10,475, of which over half, $5,800, came from the two American vessels and was probably refunded in full. The cargoes, of which the most valuable was the boatload of slaves, fetched $43,000, and so the total proceeds of sale were $53,475. The confiscated property was probably worth more. The British authorities may have exaggerated in estimating the loss at $70,000, but it is likely that Gálvez expected more than was obtained.[103] With different timing, his swoop, made shortly after a number of richly laden vessels had left for Britain, would have yielded more.

In defense of his behavior, Gálvez stood on the terms of the peace treaty of 1763. Free navigation, which the treaty allowed, did not include the right to trade with Spanish subjects. Shoreside cargoes and storehouses and even bridges built between the bank and vessels permanently moored in the Mississippi suggested the existence of British trade. To the charge that he should have arrested only on proof and not upon mere suspicion of smuggling, the governor replied simply, "I think differently."[104]

According to Martin Navarro, who, after fifteen years' service in Louisiana, urged the Spanish to adopt a free-trade policy for the province, Gálvez soon had cause to repent of his harshness toward the British. Navarro believed that such prosperity as Louisiana enjoyed during the 1760s came from the readiness of Gálvez's predecessor to overlook the illicit trade carried on by the English, who brought in settlers, provided the slaves necessary to cultivate the land, and purchased the produce of the plantations. "There was an increased industry never experienced before." He considered the results of Gálvez's action against the English to be catastrophic, "as once the importation of blacks ceased, the inhabitants began to lose the wealth that trade brings and those things which alone make for happiness and the growth of empire."[105]

With reason, then, Gálvez soon modified the expulsion order which had followed his seizure of shipping. He offered to let the British merchants return to do business in Louisiana if they promised on no account to disturb the inhabitants of the province. Some firms preferred to stay in West Florida, to which its members had fled. Others, including David Ross and Company, John Campbell, the partnership of George and Robert Ross—all of whom had suffered in the spring seizures of 1777—accepted the offer. They had some reason to believe that persecution of the English might not last. Knowing, as they must have, what had happened after Governor O'Reilly's expulsion of British merchants and his strict ban on all British trading, they expected that Gálvez too would relent—and they were right.[106] By the summer of 1777 David Ross was involved in a court case following the arrival of his vessel *Polly* from London, and in the fall George and Robert Ross employed Jean Lombard, the owner of *La Mamie,* to take a cargo of staves from New Orleans to Teste Island.[107] No doubt these firms were responsible for other voyages, but we know nothing of them because they did not give rise to court cases. In any case there was an appearance of business as usual for English traders.

They were reluctant to leave New Orleans even when marauding Americans threatened their prosperity. In the early months of 1778, a Royal Navy vessel, the *Sylph,* arrived from Pensacola in the wake of Willing's raid. Its commander, John Ferguson, sent a message to Britons living in Louisiana, offering to take them aboard and carry them to safety. Fourteen of them replied but showed little gratitude for the offer, although those with nothing to lose, like Hannah Ogilby, took advantage of it.[108] Trading concerns, they wrote, could not be wound down overnight, and slaveowners could not abandon their property. Rather than evacuation, dispatch of a hundred redcoats would be a better answer to their problem. In a separate letter to the commander, Patrick Morgan and Robert Ross, while declining to leave New Orleans, suggested that Ferguson should delay his own departure for five days so that the *Sylph* could escort to Pensacola their ship, the *Live Oak,* which was to be loaded with indigo and peltry for London.[109]

The British authorities could not have been pleased. There was the maddening possibility that these British subjects were aiding George III's enemies. The gunpowder cargo of this same *Live Oak,* which had sailed from London for New Orleans in the fall of 1777, was consigned to the firm of Morgan and Mather for sale to the Spanish commissioner for Indian affairs, ostensibly for distribution among the tribes under his jurisdiction. Governor Chester had been unable to interfere with the *Live Oak* because, so it seems from study of the provincial secretary's account book, she did not call at a West Floridian port. After his raid it seemed probable that Willing had been resupplied with powder imported from Britain. Chester was furious and suggested that the plantations secretary should ban all British exports to New Orleans.[110]

Gálvez had been ambivalent toward Britons hurt by Willing. Some had successfully sought protection on Spanish soil, for which Henry Stuart, the British deputy Indian Commissioner, had written Gálvez fulsome thanks.[111] Others had been seized by Willing's subordinates, even in New Orleans, put in irons, and placed aboard the *Rebecca,* which the Americans had captured and anchored off the town.

The British authorities asked Gálvez to return the *Rebecca.* He refused on the ground that she was not taken in Spanish waters, but he reminded Governor Chester that he had returned property, chiefly slaves, which Willing had seized on Spanish soil. They belonged to, among others, Chester himself,

Philip Livingston, Elihu Hall Bay, and Stephen Shakespear, who had sought to secure their property by transfer from the English to the Spanish side of the Mississippi. Gálvez had also restored the brig *Neptune* because it was taken on the Spanish side of the Mississippi.[112]

On 16 April 1778, Gálvez summoned all British residents of New Orleans to Government House at eleven in the morning, took out his watch, and told them that they had half an hour in which to decide whether or not they would take a stringent oath of loyalty to the Spanish. Not only would they have to swear neither directly nor indirectly to offend or conspire against the Spanish nation, but they would also have to promise to defend it and to reveal any information they might obtain of schemes against it. Those who would not take the oath would have until noon on the following day to leave Spanish territory.

Britons who took the loyalty oath included Robert Ross and John Campbell and, it seems, David Ross, who continued in business at New Orleans despite all misfortunes. He had suffered comparatively lightly from Gálvez in April 1777 but endured a much more severe loss in March 1778, when his schooner *Dispatch,* with a cargo of fifty picked slaves and 100 barrels of flour, was seized in the Willing raid. Although Gálvez reported the restoration of both vessel and cargo a few days later, it is doubtful whether either was in its pristine state.[113] Ross then resumed business in New Orleans.

One must ask how Britons could continue to make a living there, for, quite apart from the fluctuating toleration accorded them by successive governors, business life in the town had become much more competitive. In 1777, as previously noted, the French obtained the right to supply Louisiana with slaves. At the same time, the province was given new permission to trade with Yucatan and Cuba, and the export duty was reduced to 2 percent, making resource to illicit British buyers less attractive.[114] Furthermore, Gálvez had let it be known verbally that, in settling claims, creditors who were citizens of New Orleans and vicinity were to be satisfied before any foreigners were.

French entrepreneurs were slow, however, to take advantage of opportunities in Louisiana. From 1778, when France declared war on Britain, British naval vessels hovered around the mouths of the Mississippi to arrest French merchantmen.[115] So did privateers. Captain McLean's *Rover* captured a large French schooner loaded with sugar and dry goods for Louisiana in January 1779.[116] Moreover the Louisianans still needed textiles, easily the most important British export, and neither Yucatan nor Cuba could supply them. In addition, if the Louisiana market for British slaves was shrinking, another market in the area was growing rapidly. The widely praised and publicized

English lands on the Mississippi were booming in the mid-1770s, and hundreds were settling there. In 1778 Natchez and Manchac acquired the right to send representatives to the West Florida assembly. Access to these settlements was by way of New Orleans, which formed an entrepôt for goods destined for them, and it was more cost-effective to off-load at New Orleans than to press on to the scattered upriver communities. Not only did Britons at New Orleans store goods and supply new settlers, but also they seized the opportunity created when the revolution ended New England timber exports. Louisiana and West Florida could make good the deficiency. If privateers infesting the area were avoided, it was a profitable business.

One trader who decided that a living was still to be made at New Orleans was Robert Ross, a Scot with interests on both the Spanish and British sides of the Mississippi. He had come to West Florida soon after it became a British colony, in 1765 and 1766 acquiring town lots in Pensacola and paying tax on three slaves.[117] He represented Pensacola in the provincial legislature in 1768 and 1771[118] and was already trading with John Fitzpatrick at Manchac[119] and sending slaves to Louisiana from Jamaica.[120] He had then at least one vessel, the sloop *Liberty,* but subsequently acquired several more.[121] In 1772 he was granted title to 1,000 acres on the Mississippi below Natchez.[122] On 9 March of that year, he delegated the duty of collecting his debts to John Stephenson of Pensacola, probably because of his imminent departure for New Orleans, where he established his main premises.[123] He never completed the formalities necessary for full possession of another tract of 250 acres at the Devil's Swamp near the Mississippi, for which he had applied on 4 February; while he was out of the British colony, the land was occupied by someone else.[124]

At New Orleans Ross bought tobacco and indigo and sold dry goods, Jamaica rum,[125] and slaves to the Spanish inhabitants, with the permission of the governor of Louisiana.[126] He maintained his connection with Jamaica, acting on behalf of the Kingston merchants Lewbridge Bright and David Duncomb, and apparently prospered, although on occasion he was embarrassed.[127] In 1777 Spanish authorities seized property of his because he owed the comparatively small sum of $450 to a certain William Furlong who had cheated his London partner and then sought to escape retribution through residence in Louisiana. Gálvez was sympathetic to complaints against Furlong and seized all his assets.[128] In spite of the outbreak of the revolution, the American trader Joseph Calvert brought cargo for Ross from Jamaica to New Orleans.[129] Ross survived the seizure of a brig by Gálvez in 1777, when his partner, John Campbell, lost a sloop. Both suffered grievously from the Willing raid, of which Ross wrote an excellent account.[130]

Ross and Campbell took the oath required by Gálvez for continued residence in New Orleans on 16 April 1778. Less than a month later, on 14 May, they were arrested and charged with abusing the privileges they had been given by conspiring against the inhabitants of Louisiana. Ross denied any misbehavior. As far as he was concerned, obedience to the authorities of Spain, with whom the British were not then at war, did not preclude efforts to thwart American revolutionaries, with whom they were. He had no qualms about warning the inhabitants of Natchez of the impending departure of a boat with supplies for the Americans at Fort Pitt under cover of a Spanish passport. For this cause solely, according to Ross, because of Willing's complaints, he and his partner had been subjected to a "mock trial," imprisoned for fifty-five days, and fined $600, although certain proof of their guilt was lacking.

Their activities landed Ross, Campbell, and their messenger Alexander Grayden in dire trouble—especially Grayden, who was sent to a Cuban prison. At least, though, they attained their purpose of warning the British authorities that Oliver Pollock was sending supplies for the revolutionary cause up the Mississippi. Before his arrest Grayden managed to deliver verbal warning to William Dunbar at Natchez. Gálvez wanted to avoid embarrassment. He recalled to New Orleans the vessel with the supplies. In addition to their other difficulties, Ross and Campbell had to give security against a case brought against them by a surgeon named Conand who sought $6,000 for damages resulting from the recall of the vessel on which he was a passenger.[131]

Both Ross and Campbell were banished permanently from Louisiana.[132] Ross established himself in Pensacola, where he applied for land on the Pensacola River to make up for the uncultivable grant which he had formerly been given near Natchez[133] and received a government contract to help rebuild a fort at Manchac.[134] Before the end of the revolution, he retired to London.[135]

British traders who maintained more amicable relations with the Spanish included William Walker, whom we have met as a refugee from St. Vincent and who, after establishing himself as a planter in West Florida, had diversified his activities. Following Gálvez's raid in 1777 he bought premises at New Orleans, whence he voyaged back and forth to the West Indies for rum and sugar which he sold to the English settlers of the Natchez district through John Fitzpatrick.[136] Although he had once received a bounty of 1,000 acres for loyalty to the crown, he seems to have been one of several such subjects prepared to cooperate with the Spaniards.

Despite heightened tensions, normal, if illicit, trade between England and New Orleans seems to have endured up to the outbreak of war between Britain and Spain on 15 June 1779. In May the *Live Oak* again arrived in

New Orleans from London. Her captain, now Robert Nicholson, had a crew of twenty-seven. Eight of them jumped ship as soon as the *Live Oak* reached New Orleans. The remainder did their duty until they were paid and immediately followed their mates into the town, where they stayed. The rumor was that they would crew an American frigate. Nicholson had to recruit a completely new crew and was still recruiting at the time he made a deposition in New Orleans a mere four days before war broke out.[137]

That it was possible for British traders to survive commercially in wartime New Orleans is shown by the story of the Jones brothers (the same ones involved in the *Languedoc* incident), which demonstrates the economic interdependence of West Florida and Louisiana. The brothers originated in New England, but their strongest connection was with New York, where a third brother lived, Dr. John Jones, who certainly did propaganda work for them[138] and perhaps acted for them in sending the goods which the brothers received from New York.[139] They also had a connection with Jeremiah Terry of Logan, Terry, and Company of New Orleans. Perhaps it was Terry who persuaded the Jones brothers to migrate to the gulf coast, where they arrived probably in 1765. On 21 April 1766 Evan was granted a lot on Bute Street on Pensacola's eastern side.[140] In 1767 and 1768 they paid taxes on their slaves to the West Florida government[141] and from it received either separately or jointly 500 acres on the East River, 400 acres on the East Lagoon, and some islands in the Middle River.[142]

Subsequently, following a noticeable general trend to move away from the coast to richer soil inland, James acquired 1,200 acres on the Amite River in 1772, while Evan received 600 acres on Thompson's Creek. There the climate and soil suited indigo, which, thanks to the British government bounty, could be a most profitable crop: the brothers produced between fourteen and fifteen hundredweight of it in 1775. They augmented their plantation income by lending money at 8 percent to fellow West Floridians, by serving as agents for Philadelphia and New York merchants, and by trading in New Orleans.[143] In the 1760s and early 1770s, neither of the brothers seems to have been permanently established there, although both appear intermittently. In 1769, for example, Evan successfully brought suit there against Armand and Verret for payment of $343,[144] and in December 1770 Edward Mease found one of the Jones brothers there when he visited the city.[145] As merely occasional residents they employed John Fitzpatrick as their agent to conduct their business, which consisted chiefly of selling flour and cottons and of buying cattle.[146]

In the 1760s, when West Florida was in the ascendant, the Joneses' chief activities lay in that province. Evan's career does not seem to have been seriously hurt by a duel which he fought with an unpopular governor,[147] while James was honored with a seat on the provincial council. As Louisiana acquired stability in the 1770s, the interests of the Jones brothers shifted westward. James ceased to attend West Florida council meetings after December 1773 and, when it was learned that in 1778 he had taken the oath of loyalty to the Spanish, was suspended from office.[148] He had moved permanently away from Pensacola and acquired a house on Bienville Street, New Orleans.[149]

As the possibility of an Anglo-Spanish war heightened, the Jones brothers shrewdly prepared for any eventuality. While James became a Spanish subject, Evan remained British, at least until the result of the revolution was determined. When the British cause began to fail, the brothers started to sell their land in Florida: in 1778 Evan sold 400 acres near Pensacola for $400, and in May 1779 he and his brothers sold a luckless infantry sergeant half of a Pensacola town lot for $150.[150] As soon as hostilities broke out between Spain and Britain, the Jones brothers and another English inhabitant of the Amite called Bay sought and received Spanish protection at Galveztown. All had taken oaths of loyalty to Spain.[151]

Another Briton who had transferred his allegiance from the British to the Spanish king was Captain John Davis, who immigrated to West Florida in February 1778. He successfully applied for crown land there in May of 1779 and earned a precarious living trading between Mobile and Jamaica.[152] Very soon after he received his land grant, he moved to New Orleans, where he supplied British traders farther up the Mississippi with such goods as rum, French soap, linens, gunpowder, shot, and sugar. Once the war was over he resumed his voyages to Jamaica.[153]

Yet another British merchant who continued to trade in New Orleans after the outbreak of the Anglo-Spanish war was Stephen Shakespear. Until October 1775 he had been a prosperous Philadelphia merchant. Fleeing from the upheaval of revolution, he eventually arrived at West Florida in December 1776 with four slaves. Intending permanent settlement, he sent for his wife and six children, whom he had dispatched for safety to England. With misplaced optimism he established himself in 1778 at the trading center of Manchac, thus placing himself directly in the path of Willing's raiders. Willing's men seized a loaded bateau of Shakespear's at Manchac. Together with other property taken or damaged, Shakespear calculated his loss at $8,000. Subsequently he applied for and in November 1778 was granted 1,300 acres of land in West Florida, 300 of it on family right and a further 1,000 as a bounty for loyalism. It was located on the Pascagoula River far from Manchac but not far enough to escape Spanish conquest in 1779.[154] In the aftermath he moved to New Orleans, where he engaged in trade on a small scale.[155]

Of the goods supplied to British merchants to Louisiana, there seems to have been a consistent market for textiles and a fluctuating market for flour and rum, but selling slaves was probably the most profitable business. An aggravated Francisco Bouligny described the process thus: "An Englishman in Jamaica hires a ship of 150 tons for 500 pesos at the most to come to the Mississippi loaded with goods which he obtained there on credit. With twenty or thirty negroes and with the sale of the goods, he repays the capital, pays for the freight and has money left over," even without selling all the blacks.[156]

Bouligny might well have been writing about Samuel Steer, who, as we have seen, left land and debtors in Georgia for a temporary refuge in Jamaica, where he began to practice the type of commerce described by Bouligny. In 1777 he moved his base to West Florida, installing his blacks and all the utensils and implements necessary to work a plantation on George Proffit's plantation. All fell victim to James Willing in 1778, but Steer survived. He was soon back in business, although not as a planter. In November 1778 at Kingston he shipped aboard the schooner *Eleanor* cargo for Pensacola of a kind to indicate that he wanted to sell rather than settle. Although, on the eve of the war with Spain in 1779, he was granted 500 acres in West Florida, he seems to have preferred to trade in Louisiana, where he was a notable figure, seemingly with residences in both New Orleans and Kingston.[157] In 1781 John Davis brought suit against him for selling a consumptive black. In 1782 Steer sued Colonel Anthony Hutchins of the Natchez district for a considerable sum and subsequently prospered under Spanish rule.[158]

We have no information on Steer before the American Revolution, prior to which the business of supplying slaves to Louisiana had generally been organized by London entrepreneurs with agents in West Florida. Surviving fragments of the account books of Edward Codrington show that he financed regular shipments of slaves to the Mississippi in 1773 and 1774. Conflict of interest was clearly not an issue, for his chief agent was Jacob Blackwell, the official collector of customs at the port of Mobile. Another agent was James Rumsey, a Louisiana settler who also acted for the Philadelphia firm of Baynton, Wharton, and Morgan.[159] Codrington's blacks were shipped either from Charleston or Kingston and, from the sums for which insurance was paid, probably numbered no more than two dozen on each ship.

More important than Codrington was Thomas Comyn of Love Lane, Aldermanbury, London, who was in partnership with his brother Stephen and with Nicholas Donnithorne. Their agents at Pensacola were Thomas's sons, Valens Stephen Comyn and Phillips Comyn. Valens was the more active merchant. Although Phillips traded a little in textiles and rum, his main concern was his 1,900-acre plantation on the Comite River.[160] His brother Valens had an establishment in Pensacola, where he owned five town lots,[161] which he used as collateral for loans from his father's firm,[162] and a small plantation at the mouth of the East River.[163] In the late 1760s he repeatedly represented Pensacola in the West Florida legislature; in the 1770s he was on its provincial council.[164] Accepting as payment local products, particularly indigo, he supplied slaves, textiles, and ironmongery to the Louisianans through his agent at New Orleans, who, until 1769, was John Fitzpatrick and thereafter was one of the Jones brothers.[165] Although the Comyn brothers seem to have discovered how to exploit the commercial potential of the New World, by 1781 both were dead.[166]

Others who trafficked in slaves included David Ross, who for some reason ran his business separately from his brothers George and Robert, and the firm of Morgan and Mather. David Ross was one of many Britons who immigrated to West Florida, acquiring considerable interests and property there before deciding that it would be more profitable to move to the Mississippi. He received two town lots from the West Floridian government in 1766, one in Pensacola and one in the abortive township of Campbell Town, in addition to 300 acres on the East River. In 1768 he was given permission to occupy another Pensacola lot and, jointly with one John Weir, applied for 1,400 acres on the Mississippi.[167] This application demonstrated Ross's desire to move westward, but not until 23 April 1771 did he finally obtain land there—a paltry 300 acres. Unsatisfied with this amount, on 4 February 1772 Ross applied for an additional 1,500 acres there, alleging that his family had increased, consisting by then of himself, two children, an indentured servant, and five blacks. The council allowed him to purchase merely another 700 acres.

Ross at this time was in partnership with a Pensacola merchant, Arthur Strother, and they evidently obtained their merchandise from the New York firm of Hugh and Alexander Wallace. The partnership did not always prosper. When the London merchant Prideaux Selby, on whom Strother thought he could draw, protested bills of exchange used to pay the Wallaces, the result was a court case.[168]

In the 1770s Ross decided to abandon the gulf coast and the honor of intermittently representing Campbell Town in the West Florida assembly in favor of permanent residence on the Mississippi.[169] He continued to work his plantation on the river through an overseer, and he bought houses in Baton Rouge, but his main base was New Orleans, where he bought large quantities of beaver, deerskins, and indigo.[170] In return he supplied settlers up the Mississippi with an enormous variety of items, including textiles, of course, as well as shoes, saddles, soap, Chinese tea, crosscut saws, and cutlery. Not all of it was from English sources. Coffee, French flour, brandy, cottonade, and some of his sugar were not. His foreign source of supply was Cape François. His business survived seizure of his vessels by Gálvez and Willing, and soon after the governor's expulsion order, he was trading again in New Orleans. Toward the end of the revolution, Ross evidently planned to retire in his native Britain, but by 1785 he was back in New Orleans, building up a trade connection with Jamaica.[171]

Another merchant who moved west when the gold of West Florida's promise turned to pinchbeck was Patrick Morgan. He arrived in Pensacola in 1764 and wasted little time in obtaining premises in the town, receiving lots there in February 1765 and May 1766.[172] In 1768 he attained his highest official position in Pensacola, as deputy vendue-master.[173] In 1769 Morgan was granted 2,000 acres on the Mobile River.[174] During this period he was associated with another Pensacola merchant, David Hodge, and seems to have made small sums by selling goods on commission for others like the New York merchant James Thompson.[175] Like David Ross, he decided to shift his base to the Mississippi. On 23 July 1773 he ran up a debt of over £100 with the London firm of Walker and Dawson, which had considerable interests in West Florida, and on the following day he borrowed $161 from a fellow Pensacolan, John Miller, whom he promised to pay within the month at New Orleans.[176] Next he announced his intention of returning to Europe to "settle his affairs" in England and Ireland, which was perhaps a euphemism for gathering capital, for he further intended, on his return via the West Indies, to buy twenty slaves, in consideration of which he asked for a grant of 1,000 acres of land nine miles above Pointe Coupée, on the English side of the Mississippi.[177]

Following his return from Europe Morgan worked in partnership with James Mather, an established figure in New Orleans. Although he and Mather lost a vessel to Gálvez in 1777, the scale of their business seems to have been considerable. By April 1778 Philip Barbour of the Natchez district owed them $10,984, and they were renting cargo room in their 110-ton vessel *Julie* for voyages from New Orleans to the West Indies.[178] In the same year Morgan authorized sale of lots he had bought in Pensacola through his former partner, Arthur Strother, a sign that he foresaw an end to commercial prospects there.[179]

His partner, James Mather, had a long career in Louisiana. Originally a merchant in Bochin Lane in the City of London, he emigrated to New Orleans in the 1770s. Like other Louisianans of British citizenship, he acquired free land in Florida, in his case 408 acres on the Mississippi, five miles above Bayou Manchac, but he was firmly based in New Orleans. As such he often represented absent British merchants in the court of the town. He was an energetic executor of the estate of John Waugh, who died in 1777, bringing an action on his behalf against Oliver Pollock and defending the estate against the dubious claims made by John Davis.[180] Two years later he represented John Campbell; in 1783 he represented the estate of the Mobile merchant William Struthers, who had gone insane, and in 1784 he spoke for Thomas Bowker, a Mississippi planter.[181]

That Mather and Morgan traded heavily in slaves is suggested by the seizure of their vessel with twenty-two slaves aboard in 1777. Another specialty of theirs was gunpowder and ammunition, for which, on occasion, John Fitzpatrick supplied the lead.[182]

The Spanish conquest of West Florida did not damage Mather's commercial opportunities. On the contrary, it eliminated some of his British competitors, and the Spanish government allowed him, together with Morgan's old partner Arthur Strother, to supply an annual shipload of merchandise to Pensacola and Mobile. Operating on this scale proved too ambitious, and in 1785 they had to relinquish to William Panton the right to trade with Pensacola and, in 1787, with Mobile as well. Nevertheless the Mather family was by then firmly established on the Mississippi. James's son George was loyal to Spain, but his grandson George Mather Junior took the anti-Spanish side in the West Florida revolt of 1810.[183]

The traders discussed in this chapter seem to have come from every part of the British Isles and from Britain's American and West Indian colonies. It is hard to find one who did not have a substantial stake in West Florida and difficult too to discover any who wanted to stay and trade in New Orleans who were not tolerated even in time of war between Spain and Britain, provided only that they were prepared to show loyalty to the Spanish authorities and to endure the very occasional enforcement of the not very practical trade regulations for Louisiana that had been devised in Spain. This tolerance surely indicates that they were useful to Louisiana and not mere economic parasites.

In fact British trade with Spanish Louisiana in the 1760s and 1770s served the interests of the province quite well. It supplied buyers for local products for which there was little market in Spain or in those few Spanish colonies allowed to buy from Louisiana who, unlike buyers from Spain and France, would accept those local products in small quantities. It offered Louisianans in a convenient way goods like slaves and ammunition, for which there was a consistent demand in the province, and textiles of finer quality than Barcelona could produce. Finally it enabled Louisianans to obtain these goods on credit or by payment with bills at a time when specie was usually scarce.

Very occasionally Spanish governors of Louisiana observed their uncompromising instructions to allow no trade with the British. More generally, though, they recognized that the presence of British traders was necessary for Louisiana's economic health and allowed them to go about their business. Surviving evidence nevertheless suggests that Louisianans always bought more than they sold. There was a permanent balance of payments deficit which was made good with specie from Spain. Although the interests of Louisiana may therefore have been served by illicit trade with the British, Spain paid a price; since, in strategic terms, Louisiana formed a valuable buffer between the British empire and New Spain, the Spanish king may have thought the price worthwhile.

Purely economically, English trade with Spanish Louisiana served the purposes of the British better than those of the Spanish empire. The province served as an outlet for British manufactures like shoes, hats, cutlery, iron-mongery, and, above all, textiles. Whether the textiles were entirely British or, say, German textiles reexported from Britain, they benefited Britain's economy, although French items such as those David Ross imported from Saint Domingue did not. In addition Louisiana was a source of items which Britain needed and could not produce herself: deerskins, beaver pelts, tobacco, and indigo. In the case of indigo the British crown was spared the necessity of paying the bounty due if it had been grown in British colonies.

During the American Revolution, when the traditional trade between New England and the West Indies was disrupted, Louisiana provided an alternative market for sugar, rum, and slaves. Louisiana also offered opportunities to numerous Britons, like Patrick Morgan and Evan Jones, who had been disappointed by West Florida's slow development.

Inherent in this economically beneficial trade with Louisiana were strategic disadvantages for Britain. Helping the Louisianan economy to survive strengthened future enemies. Even before Spain herself formally became a belligerent, her pro-Americanism made possible the supply from British sources of war material vital to the revolutionary cause. No doubt there would have been some such activity even if Britain had not acquired the Floridas, but the proximity of British West Florida certainly caused more Britons to reside in Louisiana and increase the volume of trade with its habitants than would otherwise have been the case. Breaking into the Spanish trade world, the exciting prospect which made West Florida an attractive acquisition in the 1760s, was best achieved, so time would show, by living under the Spanish flag. Many West Floridians did so, not knowing that in 1779 their customers would turn into invaders who would destroy British dominion in their province.

5
Plantation Life

Not everyone who came to live in British West Florida wanted to be a planter, but a great many settlers acquired plantations once they were there. The offer of free land was clearly an attraction, and a nineteenth-century Mississippi governor estimated that nine-tenths of the immigrants to the province came to cultivate the soil.[1] They arrived from a variety of colonies and countries and doubtless had widely differing conceptions of how to make a living off the land, but the climate, soil, and location of the province dictated what was suitable. Robert Rea, who has written the best survey of plantations in West Florida, concluded, after considering fourteen of them, that plantations there, whether near Pensacola, Mobile, Natchez, or Manchac, showed an essential similarity and that differences between them were minor. Plantations tended to be large. They were seldom as little as a hundred acres and often consisted of several thousand acres—too big, while the seemingly perpetual manpower shortage persisted, ever to be fully cultivable—but the houses on them were modest in size. It would be a serious misconception to visualize pillared mansions like Tara of *Gone with the Wind* fame in British West Florida. The general shortage of capital and of skilled labor and the fact that the colony never outlived its infancy may explain the lack of plantation mansions. James Bruce was one of West Florida's more substantial citizens. The main building on his plantation was a one-room house measuring sixteen feet square. The walls were weatherboarded, the floors planked, the roof shingled, and it contained a fireplace, but even so it was a very simple dwelling.[2]

In contrast was the plantation house of Arthur Neil, the keeper of the provincial ordnance store, a man of equal substance in the colony. His two-storied house on the Greenfield plantation was exceptionally large by West Floridian standards. While even modest planters had a loft in the upper story of their small dwellings, Neil's second-floor garret was spacious enough to hold a dozen upholstered mahogany chairs and a table. In the main house below were a dining room, a bedroom, and two piazza rooms. A piazza was properly an open space, but eighteenth-century Americans used the word for a veranda. Such piazzas were commonplace in West Florida and in French and Spanish Louisiana, both for public and private buildings.[3] The two piazza rooms, or verandas, at Greenfield were at least half enclosed, probably because they contained a substantial amount of furniture of a type liable to damage if exposed to the elements, including, in the front piazza, two feather beds, a spinet, and five stands of militia arms. The back piazza served as a larder and contained such items as bread baskets, butter stands, and tea chests. The main kitchen, dairy, store, and workroom (whose contents suggest laundry work) were probably separate from the main house, as most certainly was the "Negro house."[4] Neil's house was also unusual in that it was built, not necessarily on stilts, but high enough off the ground that the space beneath it could house a one-horse chaise. Greenfield was located about twelve miles west of Pensacola near the West Lagoon, where Neil was granted 2,000 acres in 1767, three years after his arrival in the province with his wife Elinor and family aboard the sloop *Friendship.*[5]

Greenfield probably had one of the most civilized plantation houses in the province. Apart from the spinet, fiddle, and French, Greek, Latin, and history books listed in a surviving inventory, there were items redolent of a cultured existence such as carpets, quilts, curtains, silver-handled knives and forks, china ornaments, a gilt mirror, and thirteen pieces of yellow wallpaper.[6] The only other wallpaper known to exist in West Florida was in Lieutenant Governor Montfort Browne's plantation, New Grove, also near Pensacola, which in Browne's absence was vandalized by Indians in 1771.[7]

The evidence that the Greenfield inventory affords of a comfortable, almost luxurious, life perhaps explains why Neil was permanently in debt, but somehow he survived and was in West Florida for at least thirteen years. Maybe he spent too much on the house and too little on stocking his farm, because, at the time the inventory was made, he had only twenty-nine cattle, four horses, and assorted swine, goats, poultry, and sheep. No doubt there were crops on Greenfield but probably only enough to feed those who lived there. The small number of male slaves—seven—and the predominance of cattle, most of whom were cows, suggest that dairy farming was the principal activity on the plantation, particularly as there was a market for dairy products in nearby Pensacola. For purposes of comparison it may be noted that planters in the 1770s who were interested chiefly in raising cattle often had herds of between 200 and 1,000 head.[8] Anthony Hutchins, for example, who lived near Natchez, claimed that during the Willing raid of 1778 he lost 500 head of cattle, 100 horses, and 200 hogs,[9] and Horn Island contained an estimated

700 head of horned cattle.[10] At his plantation on the Tensaw River, Patrick Strachan alleged that he had 600 head of cattle and 80 horses, and David Hodge claimed to have similar numbers at his place on the Perdido.[11]

Apart from New Grove, Montfort Browne owned a second plantation which was in some ways comparable to Greenfield. It was located on Dauphin Island south of Mobile and was unusual in having no slaves on it, although it did have a considerable amount of livestock and poultry. There seems to have been no fencing. The numerous hogs roaming the island ate the young turkeys, but sheep and horses throve there, as did cattle. There were so many cattle that Creek and Alabama Indians commonly rowed their canoes to the island from the mainland to kill a few cattle, barbecue them, and have a party. Browne's overseer, Daniel Hickey, could not stop them. Indeed he had no alternative, when asked by armed Indians for corn, flour, and tobacco to supplement the beef they had taken, but to give them what they wanted. The number of cattle Browne had on Dauphin Island is unknown, but it may be some guide that, when the Indians had slaughtered 114 cows in addition to bullocks and calves, Hickey estimated that they had only elimi-nated the increase in stock and had not diminished the herd.[12]

A contrast to Browne, whose multifarious activities precluded permanent residence on either of his plantations, was Major Robert Farmar, one of British West Florida's most successful residents. When he died in 1780 Farmar's estate, which comprised a plantation on the Tensaw River and another on Horn Island, was appraised at over $31,000, even though he admitted to indifferent success in his attempts to grow rice and indigo.[13] Another suc-cessful planter was Elias Durnford, whose estates were allegedly worth nearly $66,000. Perhaps more successful still, since even very valuable land and livestock cannot always be made to produce high income, was Patrick Stra-chan, who from much smaller acreage claimed that he made $4,300 a year.[14]

The Pensacola and Mobile districts were better suited to pastoral farming than to agriculture, but elsewhere in the province agriculture was very im-portant. The chief crops were indigo, rice, and tobacco. Incidental industries included the production of the materials for buildings and for barrels and the preparation of naval stores.

Before the development of chemical dyes, indigo was extremely important for coloring the products of the British textile industry. Traditional sources of this commodity lay within the French and Spanish empires, but from 1748 the British government began to pay a bounty of sixpence a pound on indigo produced in British colonies. Planters in South Carolina and Georgia were quick to take advantage of this largesse, and as early as 1764 the Floridas too began to grow indigo.[15] Even after the bounty was reduced to fourpence a pound in 1771, indigo production was still profitable.

The asparagus-like plant required much attention.[16] After weeding, planters placed half a dozen indigo seeds into holes in the ground drilled five inches apart. Once the plants started growing, they had to be protected from grasshoppers. A supposedly effective deterrent was a mixture of tar and sulfur placed in pots and burned before sunrise in the fields. When the young plants were six inches high, the dirt around their roots had to be loosened. Once they were three and a half feet tall, the indigo plants were cut down and taken to a shed which contained three vats of different levels. The largest, or "rotting," tub stood at the top. It had to be three-quarters filled with cut indigo before being topped up with water. There was controversy over whether to mix stone lime or oyster-shell lime with the water or whether, like the French, to use no lime at all.[17] The green plant would be left to steep for four hours, during which time the water would turn blue. Once the indigo had rotted, workers removed a bung, and the blue water flowed into the next lowest vat, on the sides of which rested long poles. At the end of each pole was a bottomless bucket. The poles would be worked up and down for two hours to agitate the liquid—very hard work often assigned to slaves. A white bubbly scum would rise to the surface which would threaten to overflow the vat if periodically fish oil were not spread on it with a feather.

As the liquid settled, the topmost of three holes set in the side of this second vat, usually called the battery, was opened to let out the clearest, surface liquid. After further settling, more liquid was drained off through a second, lower hole. Finally, after yet another settling, the remaining muddy liquid was allowed to flow through the opening of the third and lowest hole into the lowest vat, known as the deviling, from which the now semisolid substance was shoveled out into sacks, which were hung up to dry on nails around the inside of the indigo shed. After further drying, this mud was placed in drawerlike molds two feet long and eighteen inches wide and set in the sun. After being smoothed with a trowel, the indigo in the molds hardened and was then cut into blocks with a wire. British customs records include occasional figures for indigo exported specifically from West Florida and show a general tendency toward growth: 486 pounds in 1768, 300 pounds in 1769, and 15,256 pounds in 1771.[18] For indigo exports the year 1772 was particularly good. Britain received 16,422 pounds from Pensacola and 1,196 pounds from Mobile, while another 2,130 pounds went up the coast, probably to Charleston. All was of foreign origin. So probably was most of the indigo exported from the province's western region.[19] The brig *Africa,* which left Manchac for London in January 1774, carried 14,244 pounds of indigo, of which only 170 pounds were British grown.[20]

We know little about the quality of West Floridian indigo, but indigo in East Florida was excellent in both quality and quantity. In March 1772

Governor James Grant reported that 20,000 pounds, only part of the provincial crop, had been shipped to Charleston, where the merchants had paid
nine shillings a pound for it, twice what they would give for Carolina indigo.
Grant complained that the bounty system, whereby fourpence a pound was
paid regardless of quality, was unfair. During production, he asserted, the
quality of indigo increased as it became lighter in weight. He thought it was
against the national interest to penalize the production of high quality indigo,[21] a point emphasized by the English Society for the Encouragement of
Arts, Manufactures, and Commerce, which offered a gold medal for the
British specimen of indigo which most closely equaled the Guatemalan
product.[22]

   Indigo grows well in soil that is both light and rich, such as existed in East
Florida, but which was present only in parts of West Florida. Along the coast
of West Florida, the soil was light but poor. All indigo grown in the province
of whatever quality, and it did vary, came from the region of the Mississippi
and Amite rivers.

   John Fitzpatrick, writing from his store at Manchac in 1775, told a correspondent that he had 650 pounds of fine copper indigo available at a dollar
a pound. Four years later he offered another trader 75 pounds of what he
oddly termed "very fine inferior" indigo at four and a half ryals a pound
together with 12 pounds of "very good" indigo at seven ryals, or almost a
dollar, a pound.[23] This price was nowhere near that cited by Governor Grant
for East Floridian indigo, but the dollar a pound price was good, as we may
infer from "A Gentleman of Credit" who wrote up the attractions of life on
the Mississippi in the New York press in 1773.[24]

   Fitzpatrick sometimes took indigo as payment in kind for goods, and it is
clear that at least some, perhaps most, of the indigo he handled came from
Spanish Louisiana.[25] At one place in his interesting letter book, however, he
estimated the size of the indigo crop for 1775 of some of his West Floridian
neighbors, including three hundredweights from Evan and James Jones and
a possible six hundredweights from someone whose name is hard to read
but who was probably William Dunbar.[26] For Dunbar indigo production
was subordinate to stavemaking, but he apparently felt that it was useful to
employ convalescent and women slaves in working the vats and beating out
the seeds. Other local producers of indigo were the superintendent of the
southern Indians, John Stuart, whom Fitzpatrick supplied with an "indigo
machine," and Philip Livingston, who devoted seventy acres of his estate on
the Amite to indigo.[27]

   Concerning indigo, as is so often the case in assessing the West Florida
economy, a precise estimate of production is impossible, but we can safely

assume that indigo was grown in small but significant quantities, with potential for much greater expansion. For purposes of comparison the annual crop of neighboring Spanish Louisiana in the 1780s varied between 4,000 and 6,000 hundredweights,[28] and as early as 1769 even the East Florida crop totaled 100 hundredweights and was more than thirty times as large in 1772.[29] We may assume that, irrespective of the possible bounty which was paid only for indigo exported to Britain, indigo was a profitable crop in West Florida and that much of what was actually produced there was marketed in New Orleans, whence it would find its way to Germany, Russia, and Sweden (but not to Spain, where Guatemalan indigo was preferred).

While the governor of East Florida had boasted that his province was self-sufficient in food and in that respect had done better than any colony since the foundation of the British empire, the same could not be claimed for West Florida. Although its soil was fertile and many of its inhabitants were skillful at raising crops, yet local produce was never sufficient to feed the army and all of the civilian population. It should be remembered, however, that throughout the British period the vast bulk of the British army units in West Florida were stationed in the infertile region of Pensacola and Mobile and that the army was extremely conservative. A daily ration of flour, rice, and beef, day in and day out, was standard. The first two were unobtainable in quantity on the gulf coast. Beef was to be had, but local cattle were not used for army rations for some time. Romans wrote of the Georgia traders Rea and Galphus [Galphin?] who had the contract for supplying Pensacola with beef in 1764 and 1765 and who had to drive the cattle through the Creek nation from the adjacent colony.[30]

Subsequently the contract went to John Henniker, who allegedly received it as a reward for supporting the Grenville ministry.[31] In fact there seems to have been competitive bidding for the contract between Henniker and another London merchant, Edward Codrington, the fourth son of Sir William Codrington and the brother of a member of Parliament, who eventually supplanted Henniker as the official contractor.

Henniker had offered to supply provisions to the royal forces in West Florida at the rate of fourpence three farthings a ration. At the time West Florida was a new possession. Conditions there were so little understood in London or anywhere else that the treasury had stated that, if the sum agreed should prove to be insufficient, it might be raised. Henniker found that in fact it fell far short, for several reasons. In the hot Floridian summer butter turned to oil, and meat rotted. The government storehouses at Pensacola were quite unsuitable for preserving food. Also Henniker had not realized that it would be necessary to increase his freightage and insurance costs by sending some of the food from Ireland. In addition he had not anticipated that flour, which he obtained elsewhere in North America, would be made more expensive because of the political disturbances of the 1760s. Finally, he had discovered that the primitive facilities at the gulf ports, which lacked wharves, lighters, and labor, would also increase his expenses. The result was that Henniker and his associate Sir John Major asked the treasury for sixpence for every ration supplied at the time of contract renewal. Codrington under-bid them, and so Henniker's contract expired on 12 August 1766, after which Codrington, who had bid fivepence farthing a ration, was supposed to take over.

In fact Codrington's first cargo of provisions was lost at sea, and so Henniker's agent in Pensacola continued to use up what remained of the perhaps slightly moldy flour, beef, and rice in the storehouse, in the hope that the treasury board would consider the firm's doing so an act of patriotism and award it at least the rate given to Codrington.[32]

Despite initial hardships, Codrington proved a more satisfactory contractor than Henniker and continued to supply rations for years, helped by the renting of a large airy storehouse which replaced the original store and assisted by his agents on the spot, at one time Joseph Garrow, a landholder who found it possible to supply fresh beef rather than imported Irish beef, and Jacob Blackwell, who engaged in most of the activities—government office, po-litics, land speculation, and trade—that West Floridians pursued.[33] Codring-ton himself died in 1774.[34]

The garrison continued to depend on outside sources, but civilians even-tually found that they could become self-sufficient in food in West Florida. Most of the common fruits and vegetables that Britons and Americans were accustomed to could be grown, even in Pensacola, where, noted Rufus Putnam, there were "wonderful gardens. . . producing greens of all sorts in great plenty."[35] According to Romans, however, the land from Mobile west-ward was more suitable for vegetables; he cited cabbages, carrots, turnips, radishes, onions, asparagus, artichokes, beans, and peas.[36] Some less common eatables such as peaches, oranges, and pecans were also grown. In the richer soil near the Mississippi, abundance was possible. Bacon, beef, butter, poul-try, and orchard fruits were available in quantity, and Claiborne stated that it was common to see a hundred beehives in a farmyard.[37] That 600 pounds of beeswax were exported to Britain in 1772 lends credence to this surprising assertion.[38] Prices of foodstuffs were much lower by the early 1770s than they had been in the mid-1760s. Corn would grow everywhere, but there was a traditional preference for wheat flour and wheaten bread, which were, perhaps unnecessarily, imported for the inhabitants of the gulf coast through-out the British period.

It made good sense for a planter to grow as much as possible of the food needed by his family and slaves. He saw no reason to feed the slaves expensive imported flour. If the soil allowed, he grew rice for them; otherwise he bought cheap imported cracked rice to supplement the other basics of the slave diet—corn, peas, beans, and pumpkins, all of which could be raised on his plantation. Meat was not an important element in their meals, although the hogs that ran wild in the woods and the fish of the rivers offered additional protein to those energetic enough to catch them. There is even an instance of some blacks who caught and feasted on the flesh of two alligators.[39]

The cultivation of rice was potentially important in West Florida, and there are several known cases where it was attempted.[40] The climate was right and there was plenty of low-lying land and a constant demand for it, not just as slave food, but also because its husks made excellent animal and poultry feed. It was not widely grown during the British period, however, primarily because it was labor intensive, and labor was always scarce in West Florida. So was capital, and a mill to husk rice was an expensive item. Also rice was cheaply and conveniently available from South Carolina and Georgia.

The postrevolutionary practice of extensive water use to keep down weeds, which enormously increased yields at a cost of building and maintaining dams and drainage ditches, was of course unknown to West Floridians. Constant weeding and hoeing were nevertheless necessary and employed labor which might more profitably be used elsewhere. In the Floridas in the colonial period, rice was sown in drills eighteen inches apart in the spring. The first crop was harvested in October, and a second, inferior crop was often possible.[41] Husking rice required much labor too, since the grain is covered by an outer husk and by an inner cuticle, both of which must be removed. By the mid-eighteenth century hand-operated mills were commonly used for the comparatively easy task of stripping away the outer husk. The inner husk had to be removed by hand pounding, after which sieving removed dirt and separated the whole grain from the broken grain.[42]

William Dunbar, the only West Floridian planter to describe, even in outline, rice cultivation, probably did not have the benefit of the water- or horse-operated mills which were available elsewhere.[43] In 1776 he apparently had one large rice field by the swampland of his property and a smaller patch planted "for experiment" in the old garden on his lowland and employed the bulk of his workers in hoeing when they were available. He planted and cut his rice at odd times, sowing it in July, which was beyond the date recommended by Romans. The main harvest, however, was in August and took six days. The result was evidently unimpressive, and he did not plant rice during the succeeding three years, although he tried again in 1780.[44]

It was well known that the lands of the lower Mississippi had the climate and the deep rich soil suitable for growing tobacco. On 8 April 1765 the *New York Gazette* reprinted a report from the London press that "twenty-nine thousand weight of tobacco. . . were imported to Bordeaux. . . from New Orleans where it is said the lands to the westward of that great river [Mississippi] are exceedingly favourable for the cultivation of the plant." The lands to the east of the Mississippi were suitable too, however, and by repute Natchez tobacco was finer than Virginian.[45]

Tobacco beds were prepared in the winter months, and the seed was mixed with ashes when placed in the prepared soil, which was then trampled or beaten into compactness and covered with straw or cypress bark as a safeguard against frost. In the last week of April, the tobacco plants would have about four leaves, and the best of the plants would be transplanted into a tobacco field with three feet between each plant (Romans advised six).[46] They had to be checked morning and evening for the black worm which ate the buds. When the tobacco plants were five inches tall, the soil around them was weeded, and earth was piled up like a hill around their stems, a process repeated when the plants reached a height of eighteen inches. To make the leaves grow thick and long, the top of each plant was removed when it had acquired eight or nine leaves. At the same time any green worms found on the leaves had to be destroyed. Every planter had to build a tobacco house to cure his crop. Its design had to maximize the entry of air, to minimize that of sunlight, and totally to exclude rain.

Tobacco was ripe when the leaves turned brittle. The tobacco plants were then cut as close to the ground as possible and left on the ground until their leaves became tender, after which they were hung in the tobacco house on canes laid across the building and allowed to sweat and dry.

The roots left in the ground at harvesttime produced a second crop or even a third, such was the length of the summer on the Mississippi. If, as was customary in the region, the planter intended to roll his tobacco into "carrots," he could unhang his crop when it turned yellowish, rather than waiting for it to dry completely. The leaves would be stripped from their stalks, piled in heaps, and covered with woollen cloth. After further sweating, the center rib of each leaf was removed. Then the longest and largest leaves, those selected as the coverings of the carrots, were spread out on a large cloth of coarse linen measuring about eight by twelve inches. Onto them the smaller leaves would be piled by the handful, the cloth would be rolled up and at first tied comparatively loosely to hold the tobacco in, and then each roll would be bound as tightly as could be with a thick cord. This regional method of packing tobacco differed from the practice in older colonies, where perfectly dry tobacco was wrapped not in cloth but in other leaves and then packed in hogsheads.

Tobacco was never produced in great quantity in West Florida. Certainly enough was grown for local users, and a cargo of 320 barrels went to Charleston in 1767,[47] but none was exported to Britain in the years 1762 through 1773, with the exception of 1770, when 4,062 pounds were sent to the mother country.[48] Production continued during the years of revolution. In 1777, for the first time, tobacco was grown in appreciable amounts in the Natchez district. Although some of that year's production was sent to London by Morgan and Mather, the bulk, 500 hogsheads, was sold in New Orleans.[49] In 1780 to strengthen the fortifications at Mobile, the firm of Miller, Struthers, and McGillivray supplied hogsheads full of deerskins and tobacco.[50]

A more exotic product was cochineal, which is made from the crushed bodies of certain female insects. Apart from use as a food coloring, cochineal was in particular demand in the eighteenth century to dye wool, and West Florida had the right climate and soil for it. An optimistic forecast in the *New York Gazette* of 8 April 1765 was seemingly confirmed by shipments of cochineal to London and Charleston later that year.[51]

Perhaps these cargoes were of West Indian origin, for the news in January 1766 was that a Dutch subject, a native of Surinam, had agreed to introduce the production of cochineal to West Florida on promise of a substantial reward from the local merchants. He never fulfilled the bargain, because he died, allegedly poisoned by Spanish agents. The attempt to produce cochineal was continued, and some years later a New York newspaper reported that in 1772 sixty bags of West Florida cochineal had been imported into Bristol, England.[52] If this story was accurate the bags may have been smuggled in, because there is no report of any such importation elsewhere. Romans believed, probably correctly, that East Florida was more suitable for cochineal production.[53]

In the American colonies lumbering and the production of naval stores were agricultural industries carried on by planters. West Florida was heavily forested, and in the course of clearing woodland, planters naturally created an abundance of lumber which could be used in a variety of ways. The conversion of this waste lumber into marketable products also offered a way of profitably employing slaves at times when there were few agricultural tasks to perform.

West Florida had most of the trees useful for wood products. Areas near Pensacola and Mobile abounded with the long-leaf, yellow pine suitable for clapboards and scantlings (the uprights of door frames). Cypress used for roofing shingles and poplar for planking and boats were also plentiful. In the days before containers made of cheap metal or plastic, there was a huge demand for barrels, and near the Mississippi grew the white and red, or, more properly, black oaks, from which the staves which formed the sides of barrels could be cut and shaped. Much of the work involved in making these products could be done with primitive tools like axes and wedge-shaped froes, but sawmills were also a necessity. Although they existed in West Florida from its earliest days,[54] they were few and far between, and planters sometimes organized work parties to go up the Mississippi to cut down trees, convert them into rafts, and then ride them down the river to be cut up further in the sawmills of New Orleans.[55]

The diary of William Dunbar shows that making these products was not confined to winter. He employed women as well as men in making white or black oak staves and round, flat barrel ends called headings. The resulting production, stimulated by the offer of prize items like a colorful jacket for the slave who did best, was considerable. Dunbar could count on his smallish work force, which at most numbered twelve or thirteen, to make 4,000 staves a week. July 1777 was a particularly good month: Dunbar's team produced 17,000 barrel staves and 1,000 headings.

This kind of output required careful organization. Dunbar's blacks took part in every phase of production. They felled suitable trees and used Dunbar's sawpit in fine weather, split and shaped the staves and headings indoors if need be, and then loaded the staves onto oxcarts or packhorses to carry them to the river. There they usually off-loaded onto one of the trading vessels which regularly frequented Manchac. Sometimes he had a raft built specifically to transport staves to New Orleans, where one of the several Campbells doing business there purchased Dunbar's products.[56] There was a high degree of cooperation between the planters of the Natchez district. A number of them regularly dined together, organized stavemaking competitions among themselves, and coordinated consignments of staves downriver.

In West Florida's early years regular shipments of cedar logs were sent to Charleston, Savannah, and New York, but from 1770 the West Indies increasingly dominated as a market for lumber and wood products.[57] They included pineboards, oakboards, cedar posts, and, above all, shingles for roofing and siding.[58] War conditions stimulated the demand for West Florida wood in the islands, and as late as 25 September 1779, by which time the Natchez district had been surrendered to the Spanish and Dunbar was in a New Orleans prison, a reader of the *Jamaica Mercury* could see an advertisement for a recent shipment of West Florida white oak staves and headings. There was also a smaller but regular export of timber products to Britain throughout the 1770s.[59]

The materials for barrels and buildings were not the only byproducts of West Florida's forests. Burning vegetable substances of most kinds produced potash, which, chemically, is a form of potassium carbonate and in its purest form was known as Fine American Alkali in the eighteenth century. It was used for cleaning and bleaching. A cheaper sort with more impurities was known as coarse potash. The method for making both fine and coarse potash was similar and required steepers, or tubs and vats made of pine, cypress, or white oak. The steepers were about five feet deep and had a false bottom made of a latticework of boards. A tap was inserted on the outside of the vat near the bottom, by which fluid could be drawn from the steeper and allowed to flow into a trough.

The wood to be burnt for the best potash had to be from deciduous trees, not evergreens. It was important for greater purity to remove all charred wood or charcoal from the ashes. The ashes were then placed in a steeper whose lattice had been covered with a thick layer of straw. A depression was hollowed in the straw, into which soft water was poured. After a couple of days the tap at the bottom of the steeper was opened, and the fluid of the steeper, consisting of water with dissolved salt from the ashes, flowed out. This fluid, called the ley, would then be placed in an evaporation pan. Fresh water would be added to the steeper until the ley drawn off would not float an egg, a sign that it was time to put more ashes in the steeper.

A fire would be maintained under the evaporation pan until the salt in it solidified, a process that could take twenty-four hours. If fluid ley was removed and allowed to harden, the coarser type of potash resulted, but if the ley was allowed to remain until it became so hard that it had to be cut out with a mallet and chisel, the result would be fine potash.[60]

Unlike potash, tar and pitch were made from conifers. As naval stores they qualified for government subsidy, or "bounty." West Florida was well suited to their production, in spite of the fact that pines were used there to make tar and pitch, although it was well known that a better product came from firs. Years later, in fact, when no bounty was payable because the link with the British government had long been severed, a French writer described tar and pitch production as the most lucrative industry of the region.[61]

Apart from a little cattle ranching, making tar and pitch was the principal activity on the ill-populated British north shore of Lake Pontchartrain. Until 1777, when Governor Gálvez cracked down on smugglers, provoking the British into similar enforcement, subjects of Spanish Louisiana could without interference cross the lake, make their tar and pitch, and then run back home in their bateaux.[62] The inhabitants on the British shore would have been foolish to call in officials to curb this illicit activity, since there was ample raw material for all and they themselves, banking on Spanish tolerance, habitually sold their pitch and tar in New Orleans, in addition to most of the cattle they raised. These British subjects were a mere five or six families of French blood originally from Mobile Bay.[63]

Governor Chester was well aware of the potential of wood products to enhance the prosperity of his province and was glad when the board of trade sent him information on how the Swedes manufactured tar and pitch. The prelude to this development was a complaint by two mercantile firms of London, Bridgen & Walker and Hindley & Needham, who apparently supplied naval stores to the royal dockyards. Tar and pitch from the American colonies were inferior, they alleged, and they asked Lord Hillsborough, the plantations secretary, if the British envoy at Stockholm could obtain information on how the Swedes made their superior tar and pitch.[64] The board of trade, eager as ever to ease British dependence on the Baltic for naval stores, was, by the standards of the day, prompt to oblige. The merchants' letter of complaint had been written on 27 March 1770. By 24 August of the following year, Chester had received twelve copies of an illustrated booklet in both French and English describing Swedish methods in simple language.[65]

Tar is an empyreumatic oil (that is, tasting or smelling of burnt organic matter) mixed with resin. In a tree the substance is combined with watery matter, which, to make tar, must be removed by fire. Floods and the current threw up on the shores of Lake Pontchartrain a great deal of driftwood which was suitable for tarmaking, even though it was mostly of pine, because old trees and trees which had dried out naturally produced excellent tar.

The pamphlet on Swedish tarmaking described two methods. Much the most efficient required the building of a large furnace and stove with bricks and plate iron. Since it is most unlikely that such apparatus, despite its primitive technology, was used in West Florida, only the commoner method, the one used by the Swedish country people, will be considered here.

The main element was a pit up to six feet in depth. A pipe was laid at its bottom which led out to a cauldron. Tarwood as thick as a man's arm and cut into three-foot lengths was piled above the pipe to fill the pit. The topmost logs would be fired, and earth and turf piled loosely on them would, without extinguishing the fire, achieve slow burning so that, without bursting into flame, the lower logs would get hot enough to exude tar, which would flow out from the pipe. Initially the fluid would be watery, but then it would turn into thin tar and at last into good thick tar of the required density.[66]

Compared to using a furnace this process was slow and had other disadvantages. The tar was apt to contain earth and ash, and the quantity of tar from a given load of pines would be less than if heated in a furnace. Nevertheless the method observed in use by Charles Robin on Lake Pontchartrain twenty years after its north shore ceased to be British was quite similar to the Swedish country method.[67]

Turning tar into pitch was tricky. Molten tar had to be concentrated by keeping it hot and constantly stirred. The danger of fire was so great that kettles of tar could not be placed over an open flame. In Sweden rural tarmakers kept throwing heated rocks into the kettles. Along Lake Pontchartrain, wrote Robin, they used hot cannonballs. When the process was complete and the pitch had hardened, it was chopped with axes. A traveler to West Florida in 1776 saw evidence of tarmaking to the west of Mobile Bay, where there were three vast iron kettles, each of which could hold several hundred gallons of tar.[68] The tarmaker may have been the Frenchman Charles Parent, who had a plantation there transferred to him on security for a debt in 1772,[69] and it is sure that he made tar because, when his plantation on the other side of the bay was burned, he claimed compensation for 300 tar barrels, of which 100 were full.[70]

In the early 1760s, writers assessing the economic prospects of West Florida usually mentioned tar and pitch, and as early as 1764 barrels of both were exported to Jamaica.[71] Although the extremely common incidence of leaky vessels off West Florida's coast caused paying hulls to be a chronic maintenance occupation of all who dealt with seagoing craft, tar and pitch production never became an important industry. Even if it is impossible to estimate the amount of these products smuggled to New Orleans, the paucity of inhabitants in the adjacent region severely limited production, and known exports to Britain of tar and pitch from Pensacola and Mobile were only 478 barrels in 1768 and a mere 138 in 1770. Figures for the late 1770s, however, do seem to indicate rising production.[72]

West Florida embraced what would become prime cotton land, but cotton was of minor importance there. Governor Johnstone had gloated over the quality of some cotton seen in West Florida which weighed one and a half pounds to the shrub.[73] Some was grown, presumably for local use or, as in other colonies, to be an ornamental garden shrub, but cotton exports from the province were small and fluctuating, although generally on the increase—74 bags in 1767, 5,551 pounds in 1769, 20,457 pounds in 1770, and 3,045 pounds in 1771.[74]

Cotton was planted in widely spaced rows about six feet apart, preferably in rainy weather, and matured in five months. The right climate for it existed in West Florida and, especially along the alluvial Mississippi, the right soil too, but the conditions for selling it, which would have entailed production on a sizable scale, did not. The main reason again lay in the province's perpetual labor shortage. Cleaning cotton by hand was an uncommonly lengthy business. In colonies where slaves were more numerous, they sometimes had the task as a nightly duty. Elsewhere planters had a primitive gin comprising two treadle-turned rollers, but in West Florida at least one planter, Joseph Krebs, owned a superior machine which geared two such gins to a flywheel. A single youth operating it could allegedly clean between seventy and eighty pounds of smooth-seed cotton a day. It was one of several belonging to French planters in the province.[75]

If Dunbar's experience was typical, the life of a planter in West Florida could be comfortable if he organized his work force efficiently and escaped natural calamity, but there was little money to be made from it. Few convenient markets existed for plantation products. Vegetables grew well on Dunbar's plantation near Baton Rouge, where he grew corn, rice, peas, and cabbages. Nevertheless, apart from the windfall opportunity which came his way when Spanish conquerors gave him a contract for pumpkins, vegetables probably earned him little cash.[76] Growing them was worthwhile because it saved him from buying food for his slaves. The main settlements of West Florida, Pensacola and Mobile, were not much bigger than villages and were too distant to buy Dunbar's produce. Manchac, Natchez, and Baton Rouge were smaller still: Edward Mease calculated the entire population of Natchez at sixty-four in 1771; "hamlet" would best describe such settlements.[77] Their tiny populations kept them from being worthwhile markets. New Orleans was of more promising size, but most of its food needs could more easily be supplied from closer plantations in Louisiana.

Not surprisingly, West Florida generally welcomed new immigrants, for one reason, because they came as customers. It is also not surprising that a planter like Dunbar organized his slaves' main efforts to the manufacture of cash producers—indigo and wood products (his diary records few days in which no staves or headings were produced). The markets for them were admittedly distant, but for the West Indian buyer West Florida was a closer source of wood items than the traditional one, while the expanding demand for textile dyes in Europe made indigo, which was of small bulk and high value, a very promising cash crop, especially when South Carolina, another traditional source of supply, went into rebellion.

Outside the Natchez district the unprofitability of growing food for the market was discovered by Daniel Hickey, who was overseer on Montfort Browne's plantation on Dauphin Island. Hickey hoped to make a little money by selling his surplus corn to settlers up the Mobile River but was bitterly disappointed.[78] He eventually solved his financial problems by abandoning agriculture and running a tavern at Manchac. Bernard Lintot also found it difficult to survive as an agriculturist, despite several attempts in different locations. Enticed by the prospects from growing indigo, he obtained a 1,950-acre plantation on the Amite in 1776, found the climate unhealthy, and moved nearer the Mississippi in 1777. Success again eluded him, and like Hickey, he went to Manchac, where, in 1778, he was described as "now of Manchac, a merchant." Later he decided that Spanish Louisiana offered better opportunities for a planter and in 1780 bought an existing plantation on the Acadian coast, before deciding to return to his native New England.[79]

Along the shores of the gulf coast, as along the lakes, planters produced tar and pitch to sell to the numerous vessels frequenting the coasts or for sale in New Orleans. With time for expansion of this activity, worthwhile quantities of these products could have been marketed in Britain. Given even more time, as the comparatively empty lands of West Florida filled with immigrants, land values would probably have increased significantly—something that did not happen in the British period—and selling off unused acres might have provided another source of profit to the West Florida planter.

6

The Maritime Life
of the Province

Boats were the planes, trains, cars, and trucks of eighteenth-century West Florida. It was proper that two of its governors should be sailors, for a province bounded on the west by the Mississippi River, on the southwest by huge lakes, and on the south by the Gulf of Mexico. A number of rivers running north and south, including the Big Black, Amite, Pearl, Mobile, and Alabama rivers, did not divide but united the colony. They were the means, thanks to canoes, bateaux, pirogues, and barges, of penetrating the interior. People coming to or leaving West Florida or visiting within the province nearly always used waterborne vehicles. Governor Johnstone did cause a road to be built connecting Pensacola to Mobile, and it was possible, using Indian trails, to go on foot or horseback to Georgia and East Florida; traders did use packhorse trains to bear their ill-made, tinselly wares to tribes of the back country, and Governor Chester even possessed a wheeled conveyance. But compared to voyaging, riding was a minor activity.

Ships were the link with the civilized worlds of Port Royal, New York, Philadelphia, and London. They brought books and newspapers and furniture and mirrors and also, for a time, necessary food. In the early years too, all the immigrants came by ship; later they came from the wild western regions of colonies farther north, floating down the Mississippi on rafts and barges. The goods which the colonists needed for their own use and comfort came by sea. So too did the textiles, hats, and hardware which they hoped to sell to Spanish colonists. The warships of the Royal Navy were probably the largest vessels seen at Pensacola and Mobile. On them and on the red-coated regiments brought by contracted transports depended the safety of West Florida. The arrival and departure of boats provided variety and pleasure in what could be a very static and drab existence.

The coastal traffic that was so important to West Florida in general comprised small vessels. The schooner *Pensacola Packet* was typical. Her burden was fifty-three tons. Fully loaded she might have carried 150 barrels of rice. She had two masts and a bowsprit. The mainmast carried one sail, the foremast two, and a jib ran to the bowsprit. She lacked the newly perfected chronometer, necessary for accurate navigation. Instead for timekeeping the schooner carried half-hour glasses, although perhaps her master carried a pocket watch too. Her water supply was carried in six ironbound oaken casks. In case of leaks or heavy seas, she possessed two pumps. For use in the event of wreck, but even more for unloading cargo onto beaches, was the shallow-draft moses boat.[1]

Pensacola was West Florida's preeminent port, the only safe port capable of coping with large ships between Cape Florida and the Gulf of Mexico. It gave particularly good access to the Caribbean and the Spanish colonies of Central America. Whereas the voyage from Pensacola to New York took four weeks, the gulf port was only six days' sail from Jamaica, eight from Campeche, and ten from Veracruz.[2] Even before the British acquired Pensacola, Captain Moore took his sloop *Margaret* there from Jamaica to test trading opportunities. His daring was premature, and for his pains the Spanish jailed him.[3] Pensacola was also the regular destination of the mail boats, which were some of the most regular visitors to the shores of West Florida once it became British.

Britain had begun a packet system, chiefly for communicating with Ireland, as early as the sixteenth century. By the latter part of the eighteenth century, the system was global, enabling the directors of the East India Company in London to transmit orders regularly to Fort William, Calcutta, and the plantations secretary in Whitehall to control, as best as conditions allowed, affairs in the American colonies. The shocking defeat of General Braddock in 1755 convinced the British ministry that, cost what it might, communication between Britain and her colonies had to be as good as possible for effective prosecution of war in America. The American service initially ran between Falmouth in Devon and New York, but after peace was concluded in 1763, it was, not surprisingly in view of Britain's territorial gains in America, expanded to offer better service to the southern colonies.[4]

The packets were designed for maximum speed. They carried guns for self-defense, normally between ten and twelve small-caliber cannon, mail, some passengers, and, if the recipient were an ambassador, a governor, or a general, cargo too, although initially carrying merchandise was strictly forbidden.[5] For the New York run, packets were of 200 tons, but smaller vessels were normal on more southerly routes. Almost from its acquisition British West Florida was served by packets. On 7 January 1764 the office of the postmaster general, jointly occupied at the time by Barons Trevor and Hyde, ordered the regular sailing of packets from Falmouth to Pensacola and Charleston by way of the West Indies. There were to be three packets of 140 tons, each equipped with a crew of eighteen hands. They were to voyage first to Barbados, Grenada, St. Kitts, and Jamaica, before crossing the Caribbean and the Gulf of Mexico to visit Pensacola. Then the packets would take the long route to St. Augustine in East Florida and call in at Savannah, Georgia, and Charleston before returning to England. Packets would leave Devon on this branch of the service every two months. The first was scheduled for departure on 23 February 1764, the second on 12 April, and the third on 14 June.[6] Although the postmaster general curtailed this extraordinarily lengthy itinerary slightly in December by ordering that St. Augustine and Savannah be bypassed,[7] he also found it necessary to supplement the Falmouth packets to New York and the West Indies with a third service to Charleston to sail monthly.[8] These changes helped West Florida little. Communication by packet, the most reliable system then existing, was clearly slow and hazardous, even in time of peace.

These difficulties may be seen in the case of *Grenville,* the first packet to visit West Florida. She left Falmouth, Devon, in February 1764; by June she had got only as far as Jamaica, thus arriving at Pensacola at least four months after departure from England.[9] Her career was brief. On another voyage Captain Curlett took *Grenville* from Pensacola harbor on 24 March 1765; three days later he ran her aground on the rocks of the Tortugas, a small island group southwest of the Florida Peninsula. *Grenville* broke up, but not before the thirty-eight men and women aboard had disembarked in small boats, landing on a sandy key near the wreck with a mere seven gallons of water and forty biscuits. Curlett and six of his crew left in a pinnace the next day for Havana. Arriving five days later, they procured a schooner which rescued the castaways on 6 April. All had survived, though the mail carried by *Grenville* was totally lost.[10]

Misfortune also befell a successor packet of identical name north of Havana. A hurricane struck her on 12 October 1766. It lasted a full twenty-four hours, blowing away the foresail and mainsail, filling the cabin with water, and all but sinking *Grenville.*[11] The following year both her master and her surgeon died en voyage from Jamaica to Pensacola.[12]

Packets were not exclusively for official mail, but mail for private citizens was expensive and idiosyncratic. If living in Britain the sender paid, but if a new Floridian wrote home, the recipient in the mother country paid the necessary postage.[13] The cost of a single one-page letter from London to Florida was one shilling. For purposes of comparison, a day's pay for a British army corporal was then one shilling, which could also buy ten acres of land in West Florida. Another irrational aspect of the system was that a London correspondent writing to Barbados, closer to England than Pensacola, would have to pay one shilling and sixpence for a single-sheet letter.

As well as the Falmouth packets, two regular packets shuttled between Boston and New York and the main Floridian ports. They sailed every six weeks. Another pair of small forty-five-ton packets, again apart from those from England, joined Jamaica and the Leeward Islands to Pensacola and St. Augustine.[14]

In 1765 Governor Johnstone made no adverse comment on the operation of the packets; indeed in them he saw a means, ultimately abortive, of importing textiles to Florida for resale in New Orleans.[15] In an undated report on West Florida, written probably a little later by Mobile's collector of customs, Jacob Blackwell suggested that the packets be used for transporting bullion. By using British facilities, the Spanish colonial merchants of Veracruz or Havana could save the indulto tax, payable if they used Spanish vessels to send their silver dollars to Europe. Under Blackwell's scheme, bullion cargoes would go to London not Cadiz, where would arrive instead bills of exchange from London houses for an equivalent amount less commissions. The scheme would benefit Britain's merchants, sailors, and precious metal reserves. All that was needed was Spanish confidence in the West Floridian mercantile firm handling their money, argued Blackwell.[16] There is no record that they ever acquired that confidence.

In 1767 the already long voyage of the English packets to Pensacola was further lengthened. To the numerous ports at which they already called were added Madeira, between leaving Falmouth and arrival at Barbados, and Dominica, after stopping at distant Grenada. Sometime between 1764 and 1767, moreover, Antigua was added to their ports of call.[17]

The cumbersome packet system must have seemed needlessly slow to civil servants in England, who would see from a map that packet mail from Pensacola to Charleston would travel more than twice as far as it would if taken overland. In 1768 Jacob Blackwell was quick to quash any suggestion that a pony express would work better than packet boats. "The Indians will destroy the post and the postboy," he informed the plantations secretary, who may privately have calculated that a postboy was cheaper to replace than a packet. The shores around Florida were treacherous, and besides the unlucky *Grenville*s, the *Anna Theresa* packet had been lost off the Martyrs in the Gulf of Florida only a few months before Blackwell wrote.[18]

Shortly thereafter, the packets to Pensacola were temporarily discontinued. At once it was seen that the advantages of the packet system outweighed its several defects. It was no good relying on private enterprise to supply mail with efficiency. Acting governor Montfort Browne complained that four or five mails had accumulated in Kingston, Jamaica, because of infrequent voyages from that island to Pensacola.[19] Browne was not alone in his discontent. The British government was assailed by petitions from both sides of the Atlantic to restore a more adequate service to West Florida. One, signed by seven London mercantile firms doing business with the province, was presented on 18 April 1769, and another came from a dozen West Floridian merchants trading with Britain.[20]

London officials also had complaints. The postmaster general paid two contractors £150 a voyage to make four journeys a year to Pensacola. He was not responsible for demurrage, the detention of a vessel in port beyond the normal time. If colonial governors detained packet boats, as they sometimes did, it was to the detriment of the contractor's profit and of the public good, and the postmaster general asked the plantations secretary to order colonial governors to pay £2 a day to packet boat captains whom they detained.[21] Hillsborough declined to do so on the grounds that governors had no funds for demurrage, but he did instruct governors not to detain packet boats "except in cases of absolute necessity," in which case an explanation was required.[22]

Governor Chester acknowledged this order of 16 November 1771 on 11 July 1772—the date itself a comment on the mail system's inadequacy. He explained that he had detained a packet so that he could inform London of recent Indian depredations. In passing he made general criticisms of the working of the packet system. Much, it seemed, depended on who captained a particular vessel. He praised Arthur Clarke, commanding *Diligence*, for managing four trips a year, meaning that he completed the round trip from Falmouth in the officially approved three months. By contrast, Captain Terence McDonough of the packet *Comet* took as long as twenty weeks, because he dallied to discharge private cargoes on his circuitous voyage. The governor asked for a reduction in the number of ports of call on the way to Jamaica.[23] That *Comet* was concerned with tasks other than mail delivery is proved by the will of Richard Skinner, who voyaged to West Florida from Charleston as a packet passenger taking a stock of merchandise with him.[24] Also diverting packet masters from their prime responsibility were commissions from nonresidents of Florida. Thomas Davey, at one time commander of *Diligence,* undertook the lengthy business of applying for and selling land in West Florida on behalf of a naval officer.[25]

In 1774 the system was as bad as ever. Twenty-six West Floridians complained in a petition to Chester that, while packets theoretically left England two or three months apart, in fact they left and arrived in Florida at much the same time. The result was that the first to leave carried mail, but there was none to carry in another packet leaving shortly afterward, thus negating the main purpose of the packet system. They also alleged that dawdling in the Indies caused delay. Mail for West Florida sent on the Charleston rather than the Jamaica packet would reach the province two months earlier. Chester endorsed the suggested reform, stating that mail was taking four months to come from Europe, to the detriment of both government and business.[26] Dartmouth, the plantations secretary, responded by ordering government correspondence thenceforward to be sent to Pensacola by way of the Charleston packet.[27] At the same time he was not entirely sympathetic. The merchants' petition enclosed by Chester had in part complained that the packets were never detained for their convenience. Their needs, wrote Dartmouth, were irrelevant. A packet might be delayed only in the interest of government. He had received, presumably without pleasure, a protest from the postmaster general alleging that Chester had delayed the packet *Diligence* for a fortnight at the request of merchants.[28]

The small reform authorized by Dartmouth did not last long, for war broke out later the same year. The secretary of the board of trade ordered Chester to cease sending his dispatches to London by way of rebellious Charleston, to which packets would no longer be sent, but to revert to the Jamaica route. The change, combined as it was with the disruptions of war, led to almost intolerable delays. On 9 March 1777 Chester was mortified to discover from a letter dated 6 November 1776 that the secretary had received no dispatches from Pensacola dated later than April 1776.[29] Two years later the *Grenville* packet was attacked on its way to the West Indies by an American privateer. Gunfire destroyed all the mail for West Florida, but *Grenville* survived.[30] Other packets were not so lucky. One of the several named *Diligence* was taken in June 1779, while the packet *Hillsborough* fell victim to a Baltimore privateer the following year.[31] Weather too could destroy packets. On 9 October 1778 a hurricane struck Pensacola, picking up the *Comet* packet from her moorings and depositing her as a tangle of timber and ropes in woods a mile from the shore.[32]

During the revolution, the government ceased to rely exclusively on packets, and warships such as the sloop *Druid* were used to carry official mail. Simultaneously, however, in spite of their vulnerability to weather and hostile vessels, the packets braved the hazards of Caribbean voyages almost until there was no longer a British colony in West Florida. The *Duke of Cumberland* packet managed to find its way to London from Pensacola on 21 May 1779, but by that year the incoming service had become extraordinarily inefficient.

In August, Governor Chester lamented that he had reason to believe that five batches of London mail for West Florida had accumulated in Jamaica, unable to complete the last leg of their long route because of the lack of packet boats. The latest official mail he had received was written more than eight months previously. Besides the hazards of war, Chester suggested two reasons for the deterioration in service.[33] One was an innovation in payment for packet boats. Whereas formerly £150 sterling had been paid for each voyage, packet boat owners were now paid £1,000 annually. Thus packet masters had no incentive not to loiter.[34] The other resulted from the employment of vessels as packets not specifically designed for the job—smuggling boats and what he called ram carriers. By "ram carriers" Chester probably did not mean makeshift warships equipped with a projection at the bow for use against an enemy but rather what are elsewhere referred to as ram schooners, sometimes called bald-headed schooners, vessels unequipped with topmasts and therefore comparatively slow. Since there were still proper packet boats plying between London and Jamaica at the time, he suggested that they should be ordered to prolong their voyages by calling at Pensacola before returning to England. The result, he believed, would not merely be that he received mail monthly, but that the British treasury could save £2,000 a year.[35] Sometime in the same year, but probably not because of Chester's complaints, the postmaster general reverted to the old system of paying packet owners by the voyage rather than the year. John Campbell, who commanded the British army in West Florida, was careful to let the American secretary know how much he approved.[36]

At this time, Chester's authority was being seriously questioned by West Floridians. A petition signed by 130 citizens demanded his removal in the spring of 1779, and even a royal official was ready to defy him. The case concerned *Comet,* presumably a replacement for its ill-starred predecessor with the same name which had ended among the pine trees. Chester wanted her to take his correspondence to Britain with minimal delay, although evidently *Comet* was obliged to touch at Jamaica first. Michael McDonough, master of the packet, was told to voyage along the south side of Cuba to Jamaica, but Chester's orders were overruled by John Stephenson, acting postmaster for Pensacola, of whom Chester had once thought highly. But Stephenson was also a merchant with widespread interests in London and the Mississippi, which perhaps explains why he told McDonough to take *Comet* to Jamaica via St. Augustine and Savannah. This course would take her into waters which Chester believed were infested with French and American privateers—probably he had the Gulf of Florida in mind—and he would not entrust his important dispatches to her. Understandably he asked Germain to make it clear to packet masters that the commands of the civil governor of West Florida took precedence over the postmaster's. The surviving letter in which he made this complaint was a duplicate. It was simple prudence to send multiple copies of the same letter by various channels, and Chester did so, using packets and storeships plying to loyal New York and warships escorting the annual convoy of West Florida merchant vessels to England.[37]

The packet service was disrupted by Chester's own action when he received the unwelcome news on 9 September 1779 that Britain and Spain were at war. The governor felt that Pensacola was vulnerable to conquest because the only vessel in harbor at the time was the decayed *Stork* sloop. Chester decided to detain the *Carteret* packet, which had brought the news of war with Spain, until naval reinforcements arrived from Jamaica.[38] His council thoroughly approved. Captain Gray, like McDonough, refused to accept Chester's authority over his vessel but did obey the orders of Brigadier General John Campbell—up to a point. Campbell ordered him to cruise off the gulf coast between Pensacola and the mouth of the Mississippi, which he dutifully did until 27 November. While approaching the bar of Pensacola on returning from this duty, he unaccountably crowded on all sail, headed out to sea, and was never seen again in West Florida. He carried off with him an officer, twenty-seven marines, and Pensacola's best pilot.[39] Onlookers believed, accurately, that his destination was England,[40] where no one believed his story that the sight of a strange sail in its harbor (actually the British storeship *Earl of Bathurst*) had convinced him that the Spanish had taken Pensacola.[41]

Rumors spread in Pensacola at the time that both the *Comet* and the *Diligence* packets had been captured. The military governor of West Florida urged that, if future packets succeeded in reaching Pensacola, they should not be compelled to return to England by way of Jamaica and thus have to brave the tricky Windward Passage near Cuba, which had become hostile. Instead they should return to England by way of Georgia. D'Estaing's fleet had in fact captured the two packets, but ignoring Campbell's advice, Germain trusted the postmaster general, who promised to make the Jamaica packets more efficient for the purpose of communicating with West Florida by eliminating the Leeward Islands from their ports of call. The decision was not made until April 1780. All that Chester knew in December 1779 was that the two regular packets from Jamaica were gone and that the whole aquatic area embracing the Gulf of Mexico, the Caribbean, and the Florida Strait had been made more dangerous to British ships from Spain's entry into the war. He began, therefore, to send his dispatches overland to Georgia, to be forwarded to Britain, except when, irregularly, a packet boat or its equivalent was available, as was the case on 7 August 1780, when the *Diligence* arrived. It seems again that a traditional name was perpetuated, being here applied to a sloop doing duty as a packet.[42] The governor described the overland method as "a very uncertain and precarious mode of conveyance," to which he preferred packets, despite their disadvantages. Paradoxically the letter which he sent overland on 30 October arrived before the dispatches entrusted to *Diligence* on 7 August 1780.[43] The well-named *Diligence* reappeared in Pensacola once more on 8 February 1781.[44]

The burning desire of Governor Chester, ever since the spring of 1779, when inhabitants had petitioned the king for his removal, was to clear his name by rebutting the charges against him. As a reflection of the general unreliability of the mail system in the latter stages of the war, not until 18 February 1781 did Chester find a means in which he had confidence of conveying his counterblast to the petition to Britain. It was the annual convoy of West Florida merchant vessels under armed escort.[45] Although the packet system had by this time finally ground to an unworkable halt, so too, almost, had British dominion over West Florida. Imperfect as it was, the system yet was the main method, throughout the period of British rule, of maintaining contact with the rest of the English-speaking world and was an integral part of the functioning of the colony which must be understood if we are to understand how West Floridians lived.

Then as now the coast of the Gulf of Mexico, unlike more northerly American shores, was liable to be struck suddenly and extremely destructively by hurricanes. The hurricane that blew on 22 October 1766 reveals something of shipping in Pensacola in a not especially busy season.

Blowing from east southeast, the hurricane continued from ten at night until six the next morning, when the wind shifted to west southwest and Pensacola obtained some relief. Four vessels in harbor managed to ride out the storm with varying degrees of damage. Seven vessels were lost, although it was thought that one of them, the tender from a man-of-war, was perhaps recoverable. Definitely beyond salvage were the brig *Rebecca* of Charleston, under her master, Robert Craig, and the *Bay of Honduras* brig from Jamaica, which went down with four sailors and the owner's son. In addition four sloops and a schooner were destroyed. Thus there were at least eleven vessels in port that day.

Vessels caught in the gulf also suffered. HMS *Ferret* limped back into Pensacola on 2 November. She had been totally dismasted. Five Spanish galleons on their way from Veracruz to Spain by way of Havana were driven ashore in St. Bernard's Bay, southwest of Pensacola, from which Captain Fitzherbert in HMS *Adventure* sailed to assist them on 11 November.[46]

These numbers were unusually large for the prewar period, as suggested by the fact that on 13 November 1773 no vessel was at Pensacola except the *Ferret* sloop of war.[47] There was a revival of activity during the revolution, however, and when a hurricane struck again on 9 October 1778, fourteen vessels were destroyed and at least three other vessels were in Pensacola harbor at the time.[48]

These were unusually calamitous examples, but every year brought reports of shipwreck on or off the Floridian coasts. Trading vessels from other mainland colonies inescapably had to negotiate dangerous waters on their way to the province. They had to thread the treacherous strait between the Bahama Islands and Florida Peninsula and then had to round its southern tip, avoiding the innumerable keys surrounding it. Unless they had no alternative, prudent skippers would not call at St. Augustine, for entering its harbor was notoriously tricky. Moreover, until George Gauld's widow published his cartographic surveys in 1790, there existed no accurate charts of Floridian waters.[49]

In September 1763 Richard Savery's schooner *Dublin* was lost on the Bahama Bank. Though compelled to eat rats and raw cormorant as well as whelks and palmetto cabbages from the desolate keys, the bulk of the passengers and crew survived. In the following month another schooner, the *Charming Nelly*, was lost near the Matanzas, ten miles south of St. Augustine off the east coast of Florida, and in December the schooner *Harlequin* broke up on the St. Augustine Bar.[50]

Coincidentally a different schooner called *Harlequin* ran onto the rocks at Little Abacoa, one of the Bahama Keys at the end of May 1764.[51] In the hurricane season later in the same year, the sloop *Sally* sprang a leak shortly after leaving Pensacola on her way to Philadelphia. Captain Fisher put into Cuba to effect repairs but drove her onto one of the Colorados, small islands to the north of Cape Antonio at Cuba's western end. With no rudder and five feet of water in the hold, Fisher had no option but to take to the boats, in which he and his crew reached safety on 12 November.[52] More bizarre was the loss of the sloop *Pensacola,* which was found drifting sixty miles southeast of Mobile Point fully loaded with lumber but without a soul on board. She had been blown out of the Mobile River by a violent gale on 20 December 1764.[53]

A month later the brig *Polly,* while serving as a troop transport, was wrecked on one of the Biminis, islands on the eastern side of Florida. The captain and the forty-five passengers were saved.[54] On the well-named Dry Tortugas to the west of the peninsula, the *Grenville* mail packet was wrecked in February, the sufferings of whose crew and passengers have already been described.[55]

In March 1766 the schooner *Nelly* broke up on the bar at St. Augustine.[56]

Occasionally, while shipwreck was averted, inconvenience was not. Earlier that year a brig from Pensacola bound for Boston was blown off course all the way to distant Antigua.[57] In October one of the several packets named *Grenville* was battered for twenty-four hours by hurricane winds which took away the main and foresail; the vessel was so waterlogged that the mails were ruined, but eventually she limped into Charleston harbor.[58] Less lucky in that same hurricane month of October was the sloop *Catherine*. Wrecked on Key Largo, her crew was in an open boat for seven weeks before finding succor at Fort Apalache at Apalachicola. In the following month the brig *Tryphena,* making her way to London via the Gulf of Florida, became another victim of the Bahama Bank, as did the sloop *Fanny* on 28 December.[59] Thus without counting the previously mentioned multiple losses caused by a hurricane in Pensacola on 22 October, the year 1766 saw the loss of at least four vessels in Floridian waters.

The following year proved worse. The *St. Augustine Packet* perished on the bar of St. Augustine in February.[60] In April the Pensacola-bound brig *Sally* was wrecked.[61] In June a sloop from Honduras went down near Pensacola.[62] Experience did not always ensure safety. Captain Dyer, who had sailed to West Florida several times, lost his schooner *Britannia* after leaving Pensacola at the end of September.[63]

Balthazar Kip, like Dyer, a veteran voyager, was cast away with his vessel on the West Florida coast in the first month of 1768.[64] Two months later the brigantine *Charles Town* bound for New Orleans was lost on the Bahama Bank.[65] Later that year the sloop of Captain Robert Harris came to grief at the Mosquito Inlet on Florida's east coast.[66]

Because of the frequency of their voyages, casualties were naturally high among the mail packets. The aforementioned *Anna Theresa* packet was wrecked off the Martyrs in July 1768, although the crew, mails, sails, and rigging were all saved, thanks to the fortunate arrival of a wrecker from New Providence. Calamity was common enough for salvaging wrecks to be a major occupation in the Bahamas.[67] In December a snow acting as a transport with 160 soldiers aboard was lost on the infamous bar of St. Augustine after making a safe passage from Pensacola. Their baggage was lost, but the troops were all saved.[68]

In March 1769 the schooner *Britain* was wrecked in the Gulf of Mexico en voyage to New Orleans,[69] while the *Ogeechee* was lost on the Mosquito Bar in April.[70] In the following year the snow *Florida Packet* left Madeira for Pensacola on 19 June and ran onto a rock off Hispaniola on 10 July. It was on this occasion that Governor Chester lost his (uninsured) chariot and royal portraits.[71] Later the ship *Pompey* was destroyed on Simpson's Bar near St.

Augustine,[72] and the sloop *Anne* from New York was wrecked while trying to negotiate St. Simon's Island off the coast of Georgia.[73]

The Bahama Strait claimed another victim on the night of 15 April 1771, when the brig *St. George,* with a cargo from Honduras, presumably logwood, was entirely lost. The crew escaped death but not hardship. Heading for New Providence in an open boat, they were taken up by a Spanish schooner and carried initially to Havana and then to New Orleans and then to Pensacola. They were finally brought to New York in July.[74] The crew of another vessel, which was wrecked 120 miles west of Pensacola in the winter of 1772, faced an even more severe ordeal. They were marooned on a barren and waterless island for a month before being rescued by an engineer lieutenant from Pensacola, and in April 1773 George Stewart's brig *Sally* was cast away on the Floridian coast sixty miles north of St. Augustine.[75] Another victim of the treacherous Floridian shores was the brig *Susanna,* which was wrecked near Smyrna in September.[76] Later in the year there were two more wrecks. The first was particularly tragic. The schooner *Dove,* coming from Africa, was destroyed on the coast of East Florida. Of its cargo of 100 slaves, 80 drowned, besides 2 sailors. Two weeks later a brigantine from London was wrecked on the now notorious bar of St. Augustine.[77] In the space of little more than a decade, the bar had claimed at least half a dozen vessels. In 1774 we know of no vessel lost, although the schooner *Rose* ran onto one of the Florida Keys on its way to Mobile and had to head for Jamaica for repairs.[78]

Bad weather apparently hit West Florida's gulf ports hard on 24 August 1775. The provincial secretary laconically noted the loss of the schooner *Juno* on Mobile Point,[79] but also, if a New York paper is accurate, Daniel Moore's brigantine *George* and a schooner from the Bahamas, were wrecked at Pensacola.[80]

Thereafter war added risks to existing natural dangers. Of course, war with Spain ended the assistance which in peacetime, as has been seen, was available from Cuba, and the revolution terminated the cooperation of colonies north of West Florida, which had been common in peacetime. One example involving West Florida had occurred in 1770 very shortly after Peter Chester became its governor. Also involved was the South Carolina grandee Henry Laurens, who had misguidedly entrusted his schooner, the *Brother's Endeavour,* with a cargo of cotton and shingles, to one Magnus Watson. This master had sailed from Charleston to Kingston, Jamaica, where he had sold the cargo for something over £600. He had then sailed the schooner with its crew of five, including two blacks, Jemmy Holmes and Andrew Dross, the property of Laurens, to Mobile. There the authorities, on information from a woman passenger and Holmes, seized the schooner on suspicion that the master had violated the trust placed in him and transferred the *Brother's Endeavour* to Pensacola. Laurens wrote to Chester there and received his schooner back rather the worse for wear. Watson was jailed.[81]

Watson may have assumed that, because West Florida was a new colony, the law would be slackly enforced, an assumption with some plausibility, but not in the province's gulf ports, where many royal officials resided who had very little to do. At one time books must have existed in which were formally recorded the activities of West Florida's vice-admiralty court. They have apparently not survived, but some of those activities are known from governors' correspondence and from the minutes of the West Florida council.

Throughout the British period cases were brought before the vice-admiralty court of West Florida. An early instance concerned a vessel charged with smuggling under the navigation acts. From the rival versions of Major Robert Farmar, Thomas and Christopher Miller, Jacob Blackwell, and Edmund Rush Wegg, the following occurred.[82]

On 6 March 1765 Farmar's *Little Bob,* an open coasting vessel of the type known as a drogher, commanded by Daniel Maclean, delivered a cargo of provisions and then returned from New Orleans to Mobile with a cargo of five hogsheads of French wine bought as a private venture by Maclean. She arrived at night and at once transferred the claret to the Royal Navy vessel *Charlotte,* which was already anchored there. The customs officers sent to seize *Little Bob* discovered her alongside the brigantine *New Bumper,* Captain Roger Fagg, which had gone aground in Mobile Bay some thirty miles from Mobile. From the brigantine she unloaded eighty-nine barrels of flour, which were brought back to the town on the night of 8 March. There on the following day Daniel Clark, a deputy customs collector, seized both vessels and the cargo of flour. They were prosecuted in the vice-admiralty court at Pensacola, which was presided over by Wegg. The young judge exonerated the *Little Bob* but confirmed seizure of the *New Bumper.* Asked for a writ of probable cause of seizure of the *Little Bob,* he refused. Feeling outraged and undeniably inconvenienced, Farmar brought suit against the customs officials, claiming £400 damages for injury and delay.

The trial for damages was set for January 1766, in the provincial court of common pleas, but the deputy customs collector who had seized *Little Bob* wanted the matter raised sooner, either before the court of chancery or before the governor sitting in session with his council. Wegg had ruled that *Little Bob* might not be condemned as a smuggler because, prior to her seizure, she had obtained a pass to clear Pensacola from Daniel Clark for the purpose of helping unload the vessel aground in Mobile Bay. In making the seizure, Clark was acting on the instructions of Governor Johnstone.

The impartiality of Wegg's justice is open to doubt because Farmar and the Millers chose him as their attorney in the case and because he was suspended (though afterward reinstated) from his highest legal responsibility when Governor Johnstone dismissed him from his attorney generalship. The justice of his accusers remains in doubt, for several years elapsed before they raised official complaints against him.

There is no doubt, however, that the law had been broken. All foreign wine, following passage of 4 Geo 3, c.15 (known variously as the Sugar Act, Plantation Act, and Revenue Act of 1764), which was imported into a colony was dutiable. *Little Bob* was therefore more obviously culpable than *New Bumper,* since flour for local consumption or reexport was not dutiable. Captain Fagg perhaps lacked documents but was guilty of no patent offense. If so, he was guilty of a technicality less serious than the outright smuggling of wine. Governor Johnstone in general condoned illicit trade with Louisiana as on balance in West Florida's interest. In his insistence on the letter of the law in the case of *Little Bob,* he seems to have been pursuing a personal grudge he is known to have harbored against Farmar. Daniel Clark had more than one official job, a commonplace situation in West Florida. As the provincial secretary he had issued a pass to *Little Bob* to sail to *New Bumper,* probably in the expectation that his superior, the governor, would approve. He was mistaken and then became severe with *Little Bob* on the orders from Johnstone. There was not much doubt that personal feelings had warped normal practice in the *Little Bob* case, but the regular working of the vice-admiralty court may be better seen in the subsequent case of the brig *Africa.*

Commanded by Samuel Thomas and jointly owned by the merchants Thomas Comyn and his son Phillips, the brig left London in August 1771 and arrived in Pensacola in November. On board was sundry merchandise which was to be delivered to Lieutenant John Cambel of the Royal Engineers, who lived in West Florida.[83] His cargo consisted of a bale, three trunks, a cask, three parcels, two bopes (whatever they might be), two hogsheads of porter, and a mill. All were listed on the *Africa's* bill of lading signed by Thomas, who found, on arrival in Pensacola, that one trunk had been unaccountably lost. It contained goods worth £100. Cambel asked Phillips Comyn for compensation and was refused. Cambel therefore applied to the West Florida court of vice admiralty to arrest the *Africa* and to warn all persons with an interest in her to stand ready to appear in court and asked that, if necessary to achieve satisfaction, the court should either condemn the *Africa* to be sold or that it should order Phillips Comyn to compensate him. In June 1772 the court ruled against Comyn, who decided to appeal the decision to the West Florida council, which allowed him twelve months in which to obtain a reversal of sentence from His Majesty's privy council in London. Comyn had to provide security for the possibility that, if the year elapsed, he would be obliged to pay the original penalty and, in addition, any damages sustained by Cambel as a result of the delay. These amounts were considerable, since, so Cambel alleged, Comyn's appeal compelled him to return to London from America. The result of the appeal (because of improper filing) was that the sentence of the vice-admiralty court was confirmed on 17 December 1773.[84]

The case then drew in William Clifton, a man with good reason to be wary of charges of injustice with which Cambel would soon assail him. Clifton had been attorney general of Georgia before promotion to chief justice of West Florida, where he soon ran afoul of Governor Johnstone, who, charging malpractice, suspended him in July 1766. He was at once deprived of his salary and put to the expense of voyaging to London to vindicate himself. He satisfied the earl of Shelburne, who was then responsible for colonial affairs. By royal order he was reinstated in December 1767, but after his return to West Florida in June 1768, he was always careful not to exceed his authority. After Comyn's appeal to the privy council failed, he applied to the West Florida court of common pleas, which was presided over by Clifton, for a suit of prohibition staying further proceedings, on the grounds that the court of admiralty had no jurisdiction in the case, since its facts had to be determined by a jury—which the admiralty court lacked. Clifton doubted if he could issue a writ of prohibition in a case where judgment had already been passed but was satisfied, after consulting precedents, that he could. He issued one, and the disgruntled Cambel then took his case to London, complaining of Clifton's injustice and oppression. He appears to have obtained no immediate satisfaction, but in March 1774 he obtained confirmation of the original sentence of the vice-admiralty court in London.[85]

In spite of Cambel's suit, the *Africa* was apparently not hindered from its accustomed business.[86] Although Alexander McCullagh, West Florida's provost marshal, was not one of the principals, he was also involved in the case. Somehow he and the feisty Cambel came to blows in the court. A challenge was issued, and the pair fought a pistol duel on Gage Hill, Pensacola. Neither was hurt, and with honor satisfied, the duelists resumed their former friendship.[87]

The whole of this drawn-out, troublesome, and petty affair about a lost trunk reveals that, primitive society though it might be, West Florida had the sophisticated hierarchy of courts of older colonies and that they functioned. Court officials and litigants took themselves and the legal process seriously.

Carl Ubbelohde has distinguished three main functions for vice-admiralty courts.[88] One was enforcement of British trade laws, which here is illustrated by the *Little Bob* case. The *Africa* case shows the court performing another function, adjudication in cases brought by disgruntled seamen or merchants.

In time of war easily the most common activity of the vice-admiralty courts was the third mentioned by Ubbelohde, the condemnation and disposal of prizes. The procedure upon capture of an enemy vessel was, essentially, for the court to determine whether the capture was legitimate and who was responsible for it and to arrange for the auction of the prize. The proceeds were then divided among the officers and men of the capturing crew in proportions established by the Prizes Act of 1707. There were rich pickings in the Gulf of Mexico for the naval force, small as it was in West Florida, after Spain declared war on Britain in 1779. A number of storeships and victuallers were taken. The vice-admiralty court of Pensacola disposed of them. Sometimes a dispute arose as to who should get the credit for and the money from a prize.

With Elihu Hall Bay presiding, the court convened on 2 January 1781 to consider the seizure of the Spanish schooner *El Poder de Dios.* John Buttermere, master of the sloop *Nelly,* who had taken the Spanish prize of 15 November 1780, was not equipped with letters of marque entitling him to operate as a privateer. Royal Navy officers at Pensacola, Robert Deans, James McNamara, and Timothy Kelly, commanders of the warships *Mentor, Hound,* and *Port Royal,* objected to Buttermere's receiving any prize money, claiming that he had no right as commander of a merchant vessel to take the Spanish craft unless he was attacked by her and that boats from *Hound* in sight of *Mentor* and *Port Royal* had actually towed *El Poder de Dios* into Pensacola harbor. Therefore they, not he, under statutory law, deserved the prize money from disposal of the Spanish vessel and its cargo.

Buttermere was indignant. Returning from a voyage from Jamaica to West Florida, he had sighted the enemy schooner anchored between Mobile and Mobile Point. Considering it his patriotic duty, so he alleged, to attempt her capture, he decoyed her captain, pilot, and three sailors aboard *Nelly* and then placed a pilot and three of his crew aboard the Spanish schooner so that she could be brought into Pensacola harbor. Two days later, outside its entrance, *Nelly* and *El Poder de Dios* were boarded by men of the Royal Navy. It was true, said Buttermere, that he had no commission to make prizes, but he had been instructed by *Nelly's* owners, in the event of making a capture, to hold off from a British port until he had been ashore to obtain letters of marque to give ex post facto legality to the seizure. That patriotism was not Buttermere's sole motivation was suggested by the evidence of William Gosling, the pilot that Buttermere had placed on the Spanish schooner.

Gosling alleged that, on the night after the capture, Buttermere had made the suggestion, which was rejected, that after plundering *El Poder de Dios* they should set her adrift.

Bay was in a quandary. He did not want to alienate the navy, on whom the safety of Pensacola so much depended, but he clearly thought that Buttermere's action deserved reward. Evasively he announced that he would leave a decision to the authorities in London, where there was a court of privy councillors known as the lords commissioners for prize appeals. Meanwhile the Pensacola court condemned the schooner as a legal prize. She and her cargo were sold for £1534.4.11, and the proceeds were held by the court until it was known who should have them.[89]

Governor Chester sympathized with Buttermere. Although he did not dispute the propriety of Bay's referring the matter to London, he pointed out to Lord Germain that, had it not been for Buttermere's action, a quantity of ordnance stores would have gone to the Spanish garrison at Mobile to be used against the British. He asked Lord George Germain to see that Buttermere's initiative was rewarded.[90] The eventual outcome of the case is unknown.

In time of war the role of the officers and men of the Royal Navy in defending West Florida was obvious and predictable. More surprising perhaps was their use in peacetime as enforcers of customs regulations, which, as has been noted elsewhere in this work, was a role they filled zealously, thanks to a generous parliamentary statute of 1763, to the detriment of West Floridians' chances of trading profitably with the Spanish empire. Hobbled in various ways, that trade existed rather than flourished, with the notable exception of trade with Spanish New Orleans. Even so, it never became as important as West Florida's trade with other British colonies, which waxed and waned but which in total brought hundreds of ships to the ports of West Florida.[91]

We could gain a much more precise idea than we now have of the volume of maritime traffic to and from West Florida had the registers survived of the naval officer, whose job it was to record such traffic. They have not. Instead the evidence for this traffic comes partly from one of the account books of the provincial secretary, Philip Livingston, which lists, not particularly thoroughly, the fees paid to him by incoming and departing vessels. This book spans only the years from 1774 through 1778, when Livingston left Florida. Most of the evidence concerning vessels frequenting West Florida has to be gleaned from colonial newspapers.

These two kinds of evidence indicate that, after a slow start, there was comparatively heavy maritime traffic in the mid-1760s: 93 vessels in 1764, 108 in 1765, 62 in 1766, and 63 in 1767. Thereafter traffic dropped down to an average of 30 vessels a year for the period 1768 through 1773. Another flurry of activity followed the outset of the revolution, with a high of 68 vessels arriving at or leaving West Florida in 1776. After that, while Congress continued to forbid American merchants to trade with loyal West Florida, the sea lanes to West Florida became increasingly dangerous for British ships, the number of vessels trading with the province diminished, and it became more and more common to frequent it only in convoy under the protection of an armed escort. Such an armed convoy was ordered by Secretary Lord George Germain for the vessels bound to and from England in 1776.[92] The result was that, two days before Christmas, the ships *St. Andrew* and *Marian* and the brig *Industry* were escorted into Pensacola by the armed sloop *Atalanta*, which had been detached from the Jamaica squadron for the purpose.[93] In 1779 the annual fleet from England arrived on 2 April[94] and left again for London on 4 May.[95] The following year another convoy under heavy escort arrived on 5 April, and in September the armed victuallers *Anne & Elizabeth* and *Love & Unity's Increase,* which evidently felt no need to wait for a convoy, arrived separately from England.[96] Escorted by the armed sloop *Hound,* the last annual convoy for England left Pensacola on 25 February 1781, days before besiegers arrived to end West Florida's existence as a British colony.[97]

It was natural that, in the latter stages of the revolutionary war, small vessels should have found voyaging in West Floridian waters too risky to try. The following incidents, which all occurred in the space of less than four months in 1779, illustrate the hazards.

In June the mail packet *Diligence* was taken by the French on her way to Jamaica, and her master, John Fargie, spent the next six weeks in a prison among common criminals on Cape François. On 6 August the sloop *Chance* was taken on her way from Pensacola to Jamaica. Three days later the sloop *Mississippi Pilot Boat,* going to Pensacola from St. Augustine, was taken, and on 14 August the sloop *John and Peter* was taken en voyage from Pensacola to Savannah la Mar. On 7 September the schooner *Martha* was captured on its way from Honduras to Pensacola.[98]

In the early days of West Florida, small vessels had predominated. In 1765, for instance, we know the type of eighty-six vessels which frequented West Florida: there were thirty-eight sloops, twenty-one schooners, eighteen brigs, four brigantines, two snows, one ketch, and only two merchant ships. Yet in 1781, of the five merchantmen leaving Pensacola in the last convoy, three were ships. In wartime, merchants in West Florida clearly could not dispose of their wares piecemeal. They had to save all their exports for the annual convoy, and the capacity of the vessels comprising it had therefore to be large.

The figures available for West Florida's maritime traffic are unquestionably conservative. Because officials did not live in the Natchez district of West Florida, many vessels that traveled there went unreported. Then too West Indian papers are of little help in determining how great was the volume of trade between West Florida and the islands, but it must have been considerable. A petition in 1779 estimated that "upwards of a hundred sail of ships" were employed in exporting lumber to the West Indies. It was signed by proprietors, settlers, merchants, and others connected with and interested in the prosperity of West Florida.[99] The gallimaufry of occupations represented by the signers of this petition tells something of the nature of trading in West Florida.

Everybody realized that trade was important to all in the province and that a healthy trade was linked to military strength and law and order. The petition asked for army posts on the Mississippi and for the establishment of a court of justice at Natchez. Very few West Floridians were not directly or indirectly concerned with trade. A high percentage of them traded for a livelihood; others traded as a sideline. Some owned trading vessels, and a number of them were members of the West Florida council or of its legislative assembly. It was natural that their interests should be reflected in provincial laws. These statutes make it clear that trading by sea, not with other colonies but within the limits of West Florida itself—a considerable if small-scale trade about which little is known—was a chronic concern, for the situation of droghers was persistently troublesome. Initially they had to go through the same time-wasting and expensive bureaucratic formalities as larger ocean-going vessels, even though their normal business was to ply the short distance between Pensacola and Mobile with small cargoes of provisions like corn. On 3 January 1767 the procedures for masters of vessels other than droghers were considerably tightened. Any such master who left a West Floridian port without getting written permission of the governor or who took away any passenger, slave or free, without a separate permission was liable to a £40 fine. Such permissions were not needed by masters who had deposited security of £500; presumably from this sum any charges still owed to the customs could be deducted. In practice this stipulation meant that, except in extraordinary circumstances, commanders were obliged to find the sizable £500 security. Since such requirements would tend to deter the droghers from plying their useful if small-scale trade, on the same day, 3 January, another "Act concerning Coasters" was passed by the assembly. On obtaining a six-dollar droghing pass valid for six months, the master of a drogher could enter and clear a West Floridian port for a minimal three dollars.

Problems remained. It seems that, equipped with a cheap droghing pass, vessels were used for trading outside the limits of the province and that some droghers, because the law had not made their purchase compulsory, possessed no droghing passes. It also seems that some customs officers demanded more from droghers than the law allowed. The legislative assembly tried to set all these matters right with an act to amend the existing coasters' act on 11 January 1768.

Evidently this attempt did not solve all the problems of the droghers, for a new act was passed in May 1770. Droghers were to be spared interference at forts or by customs officials by being allowed to fly a special flag to indicate possession of a droghing pass. On the other hand, they had now to post similar security to oceangoing vessels, the sum for which was now raised to £1,000, which had to be underwritten by a person of substance residing in West Florida. The new measure was apparently designed to curb the practice of debtors' sailing away from their obligations. Despite the scorn which James Bruce, the collector of customs at Pensacola, poured upon the drogher's flag as a gift to smugglers, the act might have had some salutory results had the privy council in London not vetoed it.[100]

The background to this provincial legislation was considerable pressure from London to achieve a more efficient customs service. Following the Seven Years' War, Britain's American possessions, thanks to acquisitions from former enemies, including of course West Florida, had grown enormously. New responsibilities, combined with a determination to boost revenue from America, had caused a major overhaul of the customs service, whose deficiencies had been emphasized during the war by the sale of provisions to the enemy in the Caribbean.

The aforementioned Plantation Act was intended, among other aims, to make smuggling more difficult. Reaffirming old goals, the act decreed that the colonies could import no goods directly from continental Europe, and the so-called enumerated commodities, that is, the vast bulk of colonial exports, had to be shipped first to England, whatever their final destination. Achieving these goals had always been troublesome. Now new methods to eliminate evasion were devised.

An elaborate system of bonds and certificates sought to prevent goods from American ports from going to unlawful destinations. Slack customs officers now faced new penalties. Dedicated customs officials were given protection from harassment, and the vice-admiralty court which determined the *Africa* case was part of the new system and was deliberately juryless so that local sentiment should not thwart justice.

The reformed customs system did not realize all its aims, and the British government tried again in 1767. The reorganizer was the chancellor of the exchequer, Charles Townshend, who saw that inefficiency resulted from overcentralization. Commissioners in England tried to control American trade. At his instigation an American board of five customs commissioners, headquartered in Boston, was created. The commissioners would be salaried, as opposed to living off fees, and would be backed by a full team of clerks and law officers.

This plan to improve customs collection and arrest of smugglers had certain obvious flaws, most saliently that it required a large and extensive augmentation of the bureaucracy at a time when colonists were unusually suspicious of any such activity. Another, perhaps avoidable, defect was the low caliber of the appointed officials. They had a powerful motive for enforcing laws against smugglers to the letter because of the threefold division of the proceeds of sales from the seized vessels and their cargoes. One-third went to the crown, one-third to the appropriate colonial governor, and the remaining third to the customs officers who actually made the seizure. Captures were therefore made for violation of the merest technicalities in the Plantation Act. On occasion a violation of the law was temporarily ignored to ensure a more numerous catch later when enforcement suddenly and vigorously resumed. The involvement of customs officials in the *Liberty* and *Gaspée* affairs is too well known to warrant repetition here, but mention of them serves to emphasize the role they played in converting resentment into revolution. O. M. Dickerson has even said that, if the customs commissioners appointed in 1767 had been blessed with more winning personalities, there might have been no revolution. He has also pointed out what many historians have overlooked, namely, that the fees which, from 1765, all colonial customs officers were authorized to levy for their work in addition to their salaries amounted to additional taxation of shippers and were a major grievance in some colonies.[101]

West Florida was particularly vulnerable if the spate of legislation affecting colonial commerce after 1763 was oppressive. It was a new colony which, in most respects, could not hope for self-sufficiency. To develop, it had to import great quantities of goods, partly from other colonies but mostly from Britain. With royal officials forming an unusually high proportion of the scanty population, there was apparently no escape from whatever taxes and duties Parliament and the government might choose to levy on them, as is suggested by the fact that West Florida was one of the very few colonies where the Stamp Act was obeyed. Of course if those officials wished to encourage rather than stifle the economic life of the infant colony, they could choose to shut their eyes to certain types of commercial activity.

Where customs officials had no outside income, however, since their livelihood depended largely on fees from their profession, they had a strong motive for keeping a vigilant eye on them. Despite this, although West Florida felt the weight of new laws like the Plantation Act of 1764 and the Townshend Acts of 1767, as well as those imperial administrative reforms which gave rise to what Dickerson called "customs racketeering" in colonies farther north, the load was not onerous and far from unbearable. In response to a request from London of 7 November 1770, Governor Chester submitted a list of customs officials' fees normally levied in West Florida.[102] He insisted that they were not a source of complaint among West Florida merchants, except for those paid by small coasters carrying produce between Pensacola and Mobile, whose fees for various clearances amounted at times to more than half the worth of their cargoes.[103] After all the provincial laws on the subject, the droghers were still a problem.

Examination of surviving customs returns shows that the sums collected as duties were trifling. It was quite common in West Florida's early years for London to receive nothing from the province. In November 1769 the customs commissioners in Boston listed both Pensacola and Mobile as "preventive rather than yielding [ports] as the charges [t]here generally exceeded the receipt of duties in each of them."[104] With the passage of time there was some improvement. By 1772 Pensacola yielded £106.11.3¾ and Mobile £33.5.0 in duties, with the duty on tea being a consistent earner. Tea came from Asia, but from time to time West Indies products yielded mentionable sums. In the year 1772, from the import of foreign sugar and Jamaica rum, the sum of £76.5.6¼ was collected at Pensacola and £33.5.0 at Mobile. In the same year, under the terms of the Plantations Duties Act of 1766, which had levied a duty of a penny a gallon on all molasses, whether from a foreign or a British colony, £21.10.9½ was collected at Pensacola, which implies a total of 5,169½ gallons of molasses imported in that year.[105] Still the sums collected were paltry, which may well relate to the fact that few officials were dependent on official income. One suspects that more could have been collected had the customs officials not themselves had strong mercantile interests, Jacob Black-well, a noted entrepreneur who was also the customs collector at Mobile being the egregious example, and that extensive smuggling was practiced, which, given the inlets and long, deserted beaches of Florida, was no difficult feat.

To illustrate the main expenses of a West Floridian trader, let us consider the schooner *New Orleans Packet* on a voyage made in 1770. She was square rigged, commanded by George Gibbs, and owned by the Pensacola firm of Evan and James Jones, brothers, of whom one was a councillor. The schooner had been built in New York in 1766, had a crew of fifteen, and was armed with five guns. She left Pensacola on 10 July with a cargo comprising a hogshead of muscovado sugar, fifty barrels of flour, fifty barrels of pork, and fourteen puncheons of West Indies rum. As the name of the vessel suggests, her probable destination was Spanish Louisiana, where James Jones had strong business connections. If truly destined there, her cargo would be liable to no export duties. If all or part of her cargo were sold instead to settlers or Indian traders in that part of West Florida on the eastern shore of the Mississippi adjacent to Louisiana, then again no export duty would be payable.[106] In that case, however, when the West Indies products which formed part of the cargo were first imported into Pensacola, the Jones brothers should have paid duty on them. On the muscovado sugar—the crude unrefined residue after extraction of molasses from the product of sugarcane—the current rate was four shillings per hundredweight, and on the rum ninepence a gallon.[107] In fact the Evans brothers probably paid no such duties, claiming exemption on the grounds that the goods in question were to be reexported from Pensacola. If they were actually sold at, say British Manchac or Natchez, their evasion would be undetected, since no officials were stationed there to record imports.

The principal fees to be paid by the brothers would be to the customs officers at Pensacola, the naval officer, the collector of customs, and the comptroller of customs. Before loading at one of Pensacola's six wharves, a cocket detailing the cargo, which was necessary for all exports, would have been obtained from the collector at a cost of two shillings and fourpence. Once the schooner was loaded the collector would also have demanded one shilling and tenpence for entering out the goods aboard her and another fourteen shillings for entering out the vessel. When the schooner was on the point of departure, the naval officer would have charged eight shillings and twopence for a certificate stating that his inspection had been satisfactory. A certificate would also have been obtained from the comptroller at a cost of another one shilling and tenpence, but the bulk of the comptroller's compensation was one-third of what had already been paid to the collector. In all, then, apart from the deposit of a recoverable bond of £1,000, the *New Orleans Packet* could sail after payment of the annoying but tolerable sum of thirty shillings, or about six dollars.[108]

The aforementioned case of *El Poder de Dios* demonstrates how vital the Royal Navy had become in West Florida when war with Spain broke out, but their presence, important even in peacetime, had begun to be more keenly appreciated as soon as the American Revolution had erupted and had been further accentuated in 1778 by French entry into the war. During the years of conflict there was a chronic shortage of Royal Navy vessels available for the protection of the ports of West Florida and of British merchantmen in the Gulf of Mexico or on the Mississippi. It was natural that the British government should resort to letters of marque such as Buttermere lacked to supplement their number. These were commissions authorizing the commanders of privately owned vessels to cruise in search of craft flying the enemy flag. In 1778 the admiralty in London gave Governor Chester the power to issue such letters against both American and French adversaries, but as far as we know, only one vessel ever received letters of marque.[109] Avoiding capture was probably of more interest to merchant vessels using West Florida's ports than taking prizes.

The scary irruption of American raiders down the Mississippi in the winter and spring of 1778 caused alarm, panic, and an acute awareness of West Florida's vulnerability to attack by water. The crisis caused Governor Chester to appeal to the naval and military authorities at Jamaica for reinforcement. In response Admiral Sir Peter Parker sent the frigate *Active* of twenty-eight guns and the survey schooner *Florida*, converted to combat use with eight guns, and allowed the war sloops *Hound* and *Sylph*, which were temporarily stationed in West Florida, to remain there. Another war sloop, *Stork*, was also promised. These vessels were small, and some were rotten. They were hardly adequate to defend the lengthy coastline and waterways of West Florida, but they were a great deal better than nothing. *Florida* was detached to Lakes Pontchartrain and Maurepas, while *Stork* was supposed to protect Mobile Bay. These measures came too late to prevent a bateau crewed by Willing's men from entering Mobile Bay and taking off the merchant brig *Chance*, which was laden with a cargo of staves. Lieutenant Kirkland, commanding *Florida*, was willing to try to retake the brig, which was reportedly held at Ship Island in the Gulf of Mexico, but was embarrassed by a shortage of hands. Since he had only six fit for duty, he obtained sailors from *Active* and twenty soldiers from the Pensacola garrison.[110]

Luckily the worried governor could not know that he had received the best support that he would ever get from Jamaica. Thereafter he would discover that there were too many calls on Admiral Sir Peter Parker's limited resources for West Florida, which always came low on his priority list, to be given sufficient timely naval support. Parker's instructions were to make the defense of Jamaica and the protection of the regular convoys of sugar boats from there his first objectives, and the admiral almost continuously expected invasion by the French or, after 1779, by the Spanish.[111]

A powerful fleet under the French admiral d'Estaing was stationed at nearby Haiti through most of the summer of 1779. It left in August, when Parker's spare vessels were recruited for a British landing in Central America, but d'Estaing came back in October, and with him returned Parker's fear of invasion, which, enduring throughout the winter, enabled the unharassed

Spanish to besiege Mobile in February 1780 and to capture it in March. It was natural, in the aftermath of this calamity, that the merchant vessels at Pensacola should be pressed into government service for the next expected Spanish move, an attempt on the last British stronghold in the province. The council at Pensacola placed an embargo on purely commercial voyages which ran from 31 July 1780 to 5 February 1781. All available merchant vessels were placed at the king's disposal, the biggest of which were Thomas McMin's command, *St. Andrew,* and two other ships belonging to William Robertson, the *Swift* and the *St. Juno.*[112]

To strengthen this doubtful fighting force, the war sloops *Port Royal* and *Hound* and the converted privateer *Mentor,* of twenty-four guns, finally arrived in the spring from Jamaica. The ordnance vessel *Earl of Bathurst* had brought welcome reinforcements to Pensacola together with a useful quantity of ordnance stores and Indian presents.[113] She was made battleworthy with additional cannon from the now immobile *Stork.* In October 1780 a hurricane inflicted huge damage on the Jamaica squadron in Port Royal harbor, and still worried about Jamaica's vulnerability to invasion, Parker sent no ships to Pensacola, which was soon to lose some of its existing small protection.[114] *Hound* and *Earl of Bathurst* accompanied a convoy of merchant vessels from West Florida in the last week of February 1781. The convoy was destined for England, although there were detachments from it to Charleston and Jamaica.[115]

In the new year Parker had been besieged by simultaneous requests for ships from the Leeward Islands and West Florida, and again he was too late in responding to the appeals from Pensacola. When at last a quartet of small reinforcements arrived on 11 March, it found that a Spanish armada had preempted the anchorage outside Pensacola. The naval strength thus left to protect Pensacola had diminished to *Mentor* and *Port Royal* alone. It was not enough, and the town fell to besiegers, many of whom had marched overland from Mobile and were thus invulnerable to naval action, on 8 May 1781. The fort, houses, places of business, and their contents all became the property of the new conquerors. One of them wrote that they found "many public shops supplied with as many European and Asiatic goods as could be desired" and a frigate loaded with naval stores.[116] The spoils of war would clearly have been much greater if the convoy which had comprised three transports and five merchantmen had not sailed from Pensacola on 25 February, mere days before the arrival of the Spanish.[117]

Several observations may be made about the maritime life of West Florida, some banal, but some perhaps surprising (for instance, the high volume of intercolonial traffic, even in the revolutionary year 1776). The quantity of vessels plying to and from the shores of Florida was considerable. Apart from the warships of the Royal Navy, which maintained contact with Jamaica and which had peacetime duties like surveying and catching smugglers in the Gulf of Mexico, throughout the British period a variety of packets carried mail for and from other colonies and England. In addition there were usually a number of private vessels, mostly sloops and brigs at anchor in Pensacola and Mobile, or working their way up and down the Mississippi. Annually dozens of such craft visited West Florida.

Since the natural hazards of voyaging to West Florida took their toll of a sizable percentage of the craft engaged in it, the profit involved must have been substantial, and though the merchant traffic, as opposed to the necessary official and military, traffic diminished with the onset of war, it continued while any part of the province remained British.

Despite an inadequate system of customs collection, of which a main defect was failure to maintain a customshouse or even a customs ship on the Mississippi, West Florida was not a lawless community. There was respect there for the much-maligned vice-admiralty courts, and West Florida's vice-admiralty court could act effectively. Customs revenues were not expected to be—and were not—an important source of revenue for the British exchequer. Customs were paid, however, and in general did not excite resentment, possibly because, although undoubtedly a nuisance, they were not particularly exorbitant. In the one area of customs regulation where there was hardship and complaint, the plight of the owners of droghers was seen as one which called for redress, and repeated efforts were made by the West Florida assembly to enact laws which would satisfy them.

Surviving statistics on shipping indicate that the coming revolution did not hurt the province's maritime life; the revolution actually enhanced West Florida's opportunities for profitable voyages to the West Indies. The expansion of revolution into a war which involved France and Spain contracted those opportunities and, because the war menaced Jamaica, deprived West Florida of the naval protection which otherwise might well have insulated the colony from harm.

In Pensacola and Mobile the province had two ports quite adequate to handle the growing trade, which probably would have developed much further had it remained British. The trade with the Spanish empire could well have expanded, for, despite the appearance of strength made by the victories of Gálvez, the fundamental condition of that empire in the late eighteenth century was decay while that of the British empire, proved by the resilience with which it would recover from defeat in the American Revolution, was vigor. And much of that vigor derived from the health of the

textile-dominated British economy. West Florida included much of the potentially prime cotton-growing areas of modern Alabama and Mississippi.
That potential was under-exploited by West Florida's new Spanish masters.
It would not have been, under the kind of enterprising and industrious
immigrants who began to arrive in numbers to West Florida in the early
1770s and who are exemplified by the New Englanders of the Company of
Military Adventurers.

7

The Company of Military Adventurers

The most promising of the immigration schemes that brought newcomers to West Florida was organized by the glamorously styled Company of Military Adventurers. Its members were not what Johnstone had called the overflowing scum of empire: drifters, criminals, cheapjacks, shifty land speculators, or incompetent lawyers. The Adventurers were different. In general they were men of good family, high probity, and, in many cases, some economic substance. Prepared to work hard wherever fertile land could be had, they had no particular preference for Florida. It turned out to be the most convenient settlement site, but Detroit and the Ohio were equally desired alternatives.

Energy for the company was generated by a distinguished veteran, General Phineas Lyman. Born of an old Connecticut family, Lyman was a weaver's apprentice before going to Yale, from which he graduated in 1738. In 1742 he married Eleanor Dwight, member of another old New England family. He practiced law in Suffield, at the time part of Massachusetts but later, and thanks largely to Lyman's efforts, incorporated into Connecticut. He sat in that colony's assembly first as deputy and then as an assistant.[1] His good name and position secured high military rank for him when war between Britain and France was renewed in 1755. Lyman was made a major general and commander in chief of Connecticut forces.

A chance for distinction came early. On the Lake George expedition in 1755, his superior, Sir William Johnson, was wounded. Lyman took over as the French advanced on the colonials' camp with fixed bayonets. The green Connecticut troops, tired, thirsty, and short of powder, began to fall back. The junior officers could not hold them. Lyman, by his example and exhortations, could and did, at least according to his own account. Rout changed to resistance and then to counterattack. In the upshot the French fled, leaving their commander, General Dieskau, in colonial hands.

The British government rewarded not Lyman but the absent leader Johnson for this victory with £5,000 and a baronetcy. The government was not necessarily wrong, for there was more than one version of how victory was won at Lake George.[2] For our purposes here, it is important that Connecticut men believed that Lyman deserved the credit, and he became a hero through-out the colony. He served in seven more campaigns throughout the war, which culminated in the successful siege of Havana, where Lyman com-manded all the American provincials. He was probably as highly respected as any American officer in the Seven Years' War and had worked particularly closely with the earl of Albemarle and Jeffrey Amherst, who encouraged Lyman to establish a settlement in the West with discharged Connecticut soldiers.[3]

Actually Lyman's interest in developing the West through a speculative land company predates his military career. From 1754 he was a leading light in the Susquehannah Company, which had much in common with the Military Adventurers.[4] The Susquehannah Company was formed on the eve of the Seven Years' War by 850 Connecticut men who bought lands claimed by the Six Nations Indian confederacy in northern Pennsylvania. In 1762 it sent 200 settlers to the Wyoming valley. Resentful Indians did their best to exterminate them in October 1763, and for a number of years no Connecticut people lived there. When a second party of 40 was sent there by the Susquehannah Company in 1769, it discovered that Pennsylvania now claimed the site. They stayed there as squatters. Reinforcements from Connecticut increased their number to nearly 300 before the Pennsylvanian authorities expelled them in September 1769. Returning in 1770, they were again forced out. Obstinacy finally paid off. Three settlements of Connecticut immigrants, Wilkes-Barre (1779), Westmoreland (1774), and Lackaway (1774), were firmly established, acknowledging Pennsylvanian authority but otherwise functioning as out-posts of Connecticut. Something of the same refusal to acknowledge defeat would be seen in the Military Adventurers.

In America veterans of the Seven Years' War hoped to be rewarded with free land after peace arrived, and Lyman sought to realize the hope through publicity, political lobbying, and a carefully organized veterans' association. The first known meeting of the Company of Military Adventurers was on 15 June 1763, in Hartford. A standing committee was appointed to receive dues. Each member subscribed two dollars in 1763 and three dollars in 1764.[5] The first money collected financed a trip to England for General Lyman, whose prime duty was, presumably using wartime "connexions," to obtain a substantial land grant in North America from the British government. He sailed at once. Lyman used shrewd tactics. Very few favors were conferred by any British ministry of that era without personal application to the influential figures, and delay in rewarding servicemen was traditional. Years, even decades, passed without recognition of the most meritorious claims.

In this instance, however, the government acted with uncommon dispatch. The Treaty of Paris, which ended the war, was signed in February 1763. Only eight months later, on 7 October, by royal proclamation all who had served in America during the war and were resident there became entitled to claim free land on a scale graduated according to military rank. Field officers—majors and above—could claim 5,000 acres, captains 3,000, subalterns 2,000, noncommissioned officers 200, and privates 50. In addition the governors of Quebec, East Florida, and West Florida, that is, all the newly acquired colonies on the North American mainland, were authorized to grant land to nonmilitary settlers according to family size.

In the long run it was in accordance with these latter, comparatively paltry, terms that members of Lyman's company acquired land in West Florida. He could not have known it, but the general might as well have returned to America as soon as the proclamation was published, for all the good his lingering would do those who had subsidized his trip.

In fact he tarried with, initially, good justification. Lyman wanted to establish a large settlement, perhaps even a new colony, in the American interior. The main intention of the 1763 proclamation was to prevent any such development by forbidding settlement west of the Alleghenies. The sole area exempted from this prohibition was West Florida, which, stated the proclamation, was to have the thirty-first parallel as its northern boundary; by this definition marsh, canebrakes, and sand were all that could be settled in the western part of the new province. For Lyman to delay his departure from Britain made sense. The ministry at the time was fluid: from 1763 through 1766 no less than five individuals in succession held the post of secretary of state for the southern department, whose official duties included responsibility for the American colonies. Such changes might well secure an alteration in Britain's policy toward western settlement favorable to Lyman's wishes.

It seems that the general nursed particular hopes of influencing the last of these five secretaries, the intelligent but notoriously tricky earl of Shelburne. Both before and after he took office, Lyman sent him long, informative, tendentious letters on navigating and settling the Mississippi.[6] It was unfortunate that Lyman sought his end at a time when the main object of Britain's

American policy was raising money from existing colonies rather than founding new ones, because in principle Shelburne did not oppose colonies in the
west, believing that it was useless to attempt to halt the natural tide of western
settlement.[7] At the same time the earl was fully conversant with all the
arguments against western colonies and, being aware of the strength of the
opposition, was cautious about promoting them.[8]

Another important reason for Lyman's delay concerned the terms of the
royal proclamation. They seemed generous.

> Whereas we are desirous. . . to testify our royal sense and approbation of the
> conduct and bravery of the officers and soldiers of our armies. . . we do hereby
> command . . . our governors . . . to grant without fee or reward, to such
> reduced officers as have served in North America during the late war and to
> such private soldiers as have been or shall be disbanded in America, and are
> actually resident there, and shall personally apply for the same, the following
> quantities of lands.[9]

The precise phrasing made no distinction between regular and provincial
troops. And yet colonial governors in America, as the Adventurers feared
(with good reason, as it turned out), would interpret the terms to apply only
to officers and soldiers of the regular British army. One of Lyman's duties
was to ensure that no such discrimination would invalidate the claims of
Connecticut veterans.[10] The general did his best.

> The whole body of provincial forces are distinguished from the Europeans by
> being excluded from any part of your Majesty's bounty though they have served
> during the war for much less pecuniary pay and were frequently encouraged
> by your Majesty and generals to expect a recompense in lands and were besides,
> by their educations, much the best fitted for the purpose of reducing a wilder-
> ness country to a state of cultivation and to render it profitable to themselves
> and their mother country.[11]

The king did not respond to this petition with a clear ruling. Colonial
governors varied in their attitude to provincial veterans. Some allowed them
proclamation land; others did not. This kind of discrimination seldom appears on lists of reasons for colonial irritation with Britain prior to the
revolution, but thousands of old soldiers were adversely affected, including
one George Washington who applied for land in West Florida and was refused
because he was a provincial rather than a regular field officer.[12] It was surely
one of many irritants which led many Americans to conclude that British
rule was intolerably unfair.

Despite initial failure to achieve recognition for provincial veterans, Lyman stayed to complete another of his responsibilities in London. He had been charged with obtaining the prize money due to Connecticut troops for taking part in the conquest of Havana in 1762.

The money involved promised to be considerable. First to be compensated were the sailors: £60,000 sterling were divided among some 13,000 siege participants from the Royal Navy. Individual allotments ranged from £20,000 for the senior admiral down to fourteen shillings and threepence (rather more than two-thirds of a pound) for each able seaman.

There was a similar disparity in the distribution among the soldiers on 12 April 1764. Lord Albemarle received £20,000, General Elliot £4,000, major generals £1,125, brigadier generals £321.8.6, majors £88.4.8, captains £30.8.1, subalterns £18.15.7, sergeants £1.9.5, corporals £1.2.6, and privates apparently nothing.[13]

Lyman was in England at the time but was able to obtain less than he might for his subordinates, since, in his haste to leave America, he had neglected to collect papers granting him attorney power from some living claimants and from the executors of eligible veterans killed on campaign.[14] In November 1763 Lyman advised Christopher Hargill of Rhode Island, who was preparing to undertake a mission like Lyman's, that, unless he arrived equipped with proper powers of attorney, he would obtain no money from the British government.[15]

The distribution of money in 1764 was by no means final. In subsequent years there were others, which may partly explain Lyman's lingering in London.[16] In 1771 at last it was reported that

> A fifth and final distribution of prize money to the officers and soldiers that were at the Havanna the 12th of August, 1762, will begin on the 28th of October: Those belonging to Major Gorham's corps, the Connecticut, Rhode Island, New Jersey, New York etc. on the 28th of November: The shares not then demanded will be ready to be paid on the first Thursday in every month for three years to come, by Jonathan Garton in Great Russell Street, Bloomsbury.[17]

Meanwhile Lyman had sent occasional encouraging reports to New England. The Military Adventurers met in 1764 to hear what he had achieved. Nothing had been specifically granted to them, but circumstances had improved in that the ministry had extended the northern boundary of West Florida from the thirty-first parallel to one drawn through the confluence of the Yazoo and Mississippi rivers and thus made available for settlement the lush riparian region around Natchez and Walnut Hills, the future Vicksburg. Lyman claimed this achievement for his own, but Clarence Carter suggested plausibly that land speculators closer to the administration—the earl of Eglinton and Thomas Robinson, for example—were probably more influential.[18] And it cannot be doubted that to their weight would have been added that of the newly appointed governor of West Florida. Johnstone was supposed to sail for his post on the *Tartar* in April but delayed his departure until after the boundary of his province was pushed northward.

The encouraged Adventurers promptly extended eligibility for membership of their company to New Yorkers on payment of the now normal fee of three dollars. With his expense account more solidly underwritten, Lyman stayed on in England, tirelessly petitioning the king, the board of trade, and the various secretaries of state. The tone of reports filtering back to the colonies was initially heartening. For instance, the shareholders could have read in the *New York Journal* of 21 January 1768 that "the provincial troops which served in America in the last war and hitherto have had no reward of lands, as the rest of his Majesty's troops who served there, are to have lands given them in a new government that is to be formed on the Mississippi." The same newspaper on 5 September, however, reported that "the petitions of General Lyman, and others, to the Crown for grants of lands, and establishing new governments in the interior parts of this continent, are finally set aside and rejected." As this second report hints, Lyman's ambition had grown. He had asked not only that lands granted to the company should form a new province and that he should be its governor but also, consistent with a governor's dignity, a baronet too. These demands smack of greed, but there was a more admirable side to the general's requests.

He believed that the consent of the Indians prior to new settlement was essential. With touching faith in education he advocated the founding of a college on public land in the new Mississippi colony. Indian, French, and English children would be taught at public expense. Agricultural labor would be a mandatory part of the curriculum and would have a permanently good effect, believed Lyman, on the outlook of the pupils.[19] Unoriginal and simpleminded though they might be, these words suggest an idealism worthy of notice for its rarity elsewhere in the annals of West Florida.

Lyman knew several men of influence in London, including Benjamin Franklin and Maurice Morgan, Shelburne's undersecretary when the earl was secretary for the southern department.[20] When Shelburne shed this responsibility, Lyman had to wage a new campaign of persuasion from 20 January 1768 with the earl of Hillsborough, who was appointed to the newly created position of secretary for the plantations. The lobbyist did not lose heart. "He still has good hopes," commented an acquaintance two days after Hillsborough took office.[21] Evidently the new secretary did not initially reveal the inveterate opposition to western settlement that would soon become apparent. When, later in 1768, Lyman's plan was thwarted again, he blamed his old comrade in arms, Sir William Johnson, who had become superintendent of Indians in the northern colonies.[22] The general still lingered in England, in 1769 meeting Joseph Trumbull, a London representative of the rising faction in Connecticut politics, who wished him well.[23] At this time, momentously, Lyman met Peter Chester, who would assume the governorship of West Florida in 1770. Thereafter Lyman ceased to concern himself with the Ohio and concentrated exclusively on lands within Chester's jurisdiction, although he may not have abandoned the idea of establishing a new colony in them. Chester consistently encouraged immigration to his province and offered special terms to Lyman, surely with board of trade approval. Chester would allow each family of Adventurers 2,000 acres.[24] On 9 February 1770 Lyman petitioned for 150,000 acres around Natchez.[25] On the tenth the political kaleidoscope received another shake, and Lord North formed a ministry. North did nothing for Lyman's company but did approve a personal grant of 20,000 acres in one continuous block in any vacant part of West Florida that the general chose.

The conditions attached to this grant were only a little unusual. It was to be peopled with foreign Protestants or with settlers from other colonies in the proportion of one to every hundred acres. If one-third of the grant were not settled within three years, the whole tract would be forfeit to the crown. After ten years, any of the land that remained vacant would revert to the king. A quitrent of a half penny an acre would become payable on half the grant after five years and on the entire grant after ten years. Any part of the grant suitable for forts, wharves, or naval yards would be reserved for crown use, as would any land suitable for mining gold, silver, copper, lead, or coal. If any of the land was suitable for growing flax or hemp, then at least one acre in every thousand should be devoted by Lyman to that purpose.[26]

Lyman would not have been surprised at these requirements. Stringent though they were, he must have been aware that they were seldom enforced. It would have been understandable if the veteran, having thus acquired for himself a substantial reward, had abandoned, after years of fruitless lobbying, his efforts on behalf of the Military Adventurers. Instead he lingered, perhaps tempted by the enormous fountain of favorable publicity for a new Mississippi colony which began to gush in the press on both sides of the Atlantic in the early 1770s.[27]

Among the more assiduous advocates of a new colony on the Mississippi was a man who aspired to its governorship. Montfort Browne, when acting governor of West Florida, had shown himself to be ignorant in administration and a petty tyrant. Nevertheless, he had some unusual advantages. He knew the country, owned large properties on the Mississippi, and could speak Choctaw, the language of the chief tribe in the area of the proposed new colony. Jointly with the earl of Eglinton and William Tayler, who had been the general commanding the British forces in West Florida, Browne submitted a vaguely worded petition to the board of trade, which read it on 10 January 1772. Browne and Tayler were promptly summoned to explain their petition and to answer questions about the mysterious Mississippi region. A year later the board replied to the petition with an apparently reasoned recommendation that settlement on the Mississippi should be encouraged, although creation of a new government there should not.[28] In fact reason had little to do with the board's report. The outcome of the question in no way depended on what was good for America or its colonists or even for British imperial interests. It had become a tactical issue in domestic political warfare.

It had eventually become apparent to Lyman and others that the chief obstacle to the establishment of a Mississippi colony was the Irish peer Hillsborough, who had retained his post as the plantations secretary even after Lord North became head of the government in 1770. Newspapers reported that the secretary had sworn to resign if ever the scheme for an interior colony were accepted. Since his resignation was a prime object of the powerful Bedford faction—which wanted his fall for reasons unconnected with western expansion—it supported Lyman's efforts.

The Bedford group was named after John Russell, the fourth duke of Bedford. Leading members included the duke of Grafton, Earl Gower, and Lord Weymouth. They had been temporarily important in government in 1768, but their position eroded following the accession of Lord North. They wanted to retrieve it, and to replace Hillsborough by a Bedfordite was part of their plans.[29] Columnist gossip to this effect was largely verified and elaborated by an insider, William Knox, who was undersecretary for the colonial department. The other secretaries of state, he wrote, were heavily involved in applying pressure to Hillsborough. The earl of Suffolk, who was secretary for the northern department and actually a Grenvillite allied with the Bedfords, saw some benefit to himself in discord, while the Earl of Rochford, the secretary for the southern department, wanted Hillsborough out partly because he wanted to see abolition of the office of American (or plantations) secretary (thus enhancing the importance of his own post) and partly because he had been bribed with a share in the Ohio Land Company, which sought the same sort of concessions as the Military Adventurers. The

Lord President of the Council, Earl Gower, assisted too because he aimed at the premiership and conceived that, in falling, Hillsborough might bring down North with him.[30]

Whether Knox got it quite right or not, it would seem that the efforts of many prominent politicians on behalf of an interior colony, including probably the subsidy of newspaper articles, were inspired less by imperial enthusiasm than desire for additional strength in the cabinet. For a time Hillsborough obstinately refused either to resign or to accept any of the various schemes for a colony in the North American interior. His objections were widely discussed.

One malicious and farfetched speculation was that his motives were entirely selfish. He had great estates in Ireland and had raised his tenants' rents to unreasonable heights. If the schemes for opening vast tracts on the Ohio or Mississippi were to succeed, his tenants would all emigrate to America.[31] Another, threefold analysis had more plausibility. It was hazarded that he genuinely believed that further emigration would hurt the British economy in general. Second, the difficulty of ascending the Mississippi was so great that a voyage to the new settlement would take a tedious three months, making communication with the mother country difficult and impeding effective trade. Last, so primitive was the area of intended settlement that Hillsborough thought that it stood more in need of a company of foresters than a civil establishment.[32]

Hillsborough finally resigned in August 1772 because he could not reconcile himself to an interior settlement which, to use his own words, "all the world approves."[33] At least part of the world, however, in fact supported his opinion.

Lord Barrington, for example, who was continuously secretary at war from 1765 to 1778, was convinced that "no trade worth having can now be carried on by the Mississippi."[34] James Wright, governor of Georgia, strongly opposed colonizers in the interior because "I apprehend they will soon become a kind of separate and independent people," an interesting but by no means unusual observation to be made five years before the revolution. Governor Peter Chester of West Florida was less negative. He wanted settlers on the Mississippi only because Pensacola and Mobile would benefit economically. Another reason, which Chester may have thought best left unsaid, was that they would enhance his own authority. His approval was guarded. No large grants should be given to single individuals, and he indicated a strong preference for settlers within the existing boundaries of his own province to pioneers of a separate inland colony.[35]

Some consideration may have been given to the views of colonial governors on this subject, but as a class they had been known to give very trivial opinions at times. Of undoubted weight with the government, however, were the judgments of Thomas Gage, who commanded all the British troops in North America. And his opinion, based on wide experience, knowledge, and responsibility, was that it was in the British interest to "keep the settlers within reach of the sea-coast as long as we can."[36]

The interior colony over which Hillsborough resigned was not Lyman's but one on the Ohio sponsored by the Grand Ohio, or Vandalia, Company, the largest, best organized, and most powerfully backed of all the various companies pressing the London government for land grants in the postwar years. The list of its promoters makes Lyman's military friends seem by comparison of peripheral political influence.[37] Known as the Walpole Associates, the principals numbered fifty-four and included the banker Thomas Walpole; the former governor of Massachusetts, Thomas Pownall; Grey Cooper, a joint treasury secretary; Lord Rochford, the secretary for the northern department; Lord President of the Council Earl Gower; Lord Camden, who was Lord Chancellor until 1770; Thomas Pitt, brother of the great prime minister; and John Robinson, one of the very brightest of the up-and-coming ministry men. American backers included Sir William Johnson; Joseph Galloway, the leading Pennsylvania politician; Samuel Wharton, a leading Pennsylvania merchant; and Benjamin Franklin.[38]

If the Vandalia Company were to get its grant confirmed after Hillsborough's resignation—and the prospect was excellent—the chances would apparently be good too for the Military Adventurers, since the area they wanted was much smaller than and distant from the Walpole Associates' grant. If the fall of Hillsborough had been followed by swift action, no doubt Vandalia would have come into being. Instead there were the usual administrative delays, at a time when the relation between mother country and existing colonies was increasingly strained. The alienation of that leading Vandalian, Benjamin Franklin, from the British ministry doubtless played a major role in blocking the Vandalia plan.[39]

But Lyman had not waited to see this outcome. Once Hillsborough fell from power, the general lost no time in returning to Connecticut. The story in his family was that his mind had become muddled and that he would have stayed on endlessly in England had not his wife, Eleanor, dispatched their son Gamaliel to London to persuade his father to come home.[40] On the other hand, two anecdotes appearing in the contemporary press indicated that the general remained in England so long not because his mind was disordered but because he was too poor to pay either his debts or the price of a passage back to Connecticut.

According to one story, the earl of Hillsborough feigned solicitude, with the intention of humiliating Lyman. He said that a number of the general's old friends, Lord Albemarle, General Amherst, General Monckton, and Admiral Keppel, had all agreed to subscribe £100 each to alleviate his distress and, declaring that he hoped Lyman would allow him also to be numbered with them, pressed £100 on the Connecticut veteran. When, with misgivings, Lyman had the temerity to mention the subject to his other supposed benefactors and found that they knew nothing of the scheme, he had already spent the sum given him by Hillsborough and could not even afford the gesture of returning it. Since this tale, as it stood, suggested at least some generosity by Hillsborough, its narrator, "Atticus," suggested that probably the donated £100 came from secret service funds.[41]

According to another anecdote, Hugh Boscawen, second Viscount Falmouth, was intrigued by the sight of an impoverished but dignified and gentlemanly figure at court and followed him to his home in Chelsea. Upon inquiry, he discovered that the man was Lyman and, considering him worthy of help, secretly arranged for him to be given a sum of money. He also used his own influence with the king to secure favorable consideration for Lyman personally.[42]

Lyman seems to have sailed from England after Hillsborough had fallen but without definite knowledge of his successor's intentions. He perhaps thought that the new plantations secretary, North's half brother, the earl of Dartmouth, would support western colonization, for he arrived back in America fully confident. There may have been some justification for this attitude, as suggested by a report from London of 11 November 1772 which appeared in the *Maryland Gazette* of 25 February 1773: "The settlement on the banks of the Mississippi which was so dissonant to Lord Hillsborough's system of politics is now *sub judice* before the Board of Trade, and from the warm manner it is patronized by Lord Dartmouth, there is the strongest possibility of its taking place very shortly."

Actually, it is possible that Lyman, without sure knowledge of what Dartmouth's decision would be, had made up his mind that the time had come for action, for he had talked long before, in 1767, to Benjamin Franklin of leading emigrants into the wilderness, even if he failed to obtain a grant.[43] He may have been influenced too by the time limitation on his mandamus grant of 20,000 acres. If a third of it were not settled within three years, the entire grant was theoretically forfeit to the crown. The three years were due to end on 15 June 1773.

On his arrival in Connecticut the returned hero joyfully announced that the crown had granted the company a large tract of land, allegedly of 150,000 acres. Whether he was sincere or not we shall never know. He offered no documentary proof for the assertion. The press printed the following notice:

Whereas General Lyman (after soliciting a number of years at the court of Great Britain, for a grant of land for the Company of Military Adventurers, so called) has at length obtained of the crown, a grant of a considerable tract of land, bounded west on the River Mississippi, north by the River Yasou, between the latitudes of 32 and 34; and is now, after a number of years absence, returned home; and finding by his long absence and the various disposals of Divine Providence, that several of the standing committee of said company, as also their clerk, are dead; and some of those that are yet living, being absent, therefore we the subscribers being desirous to fill up the vacancies of said committee, clerk etc., as well as to do any other business, which said committee shall think proper when they meet, more especially to do as follows, viz.

1st. To choose a president or moderator of said meeting.
2d. To choose a committee of twelve men, to explore, reconnoiter, view etc., the land granted as aforesaid on the River Mississippi.
3d. To choose a committee to call on all (by advertisement or otherwise) that have heretofore collected and received from subscribers of the aforesaid company any sum or sums of money belonging to said company; and that those who have hitherto neglected so to do, be directed to bring in the same to the treasurer of the said company; and that all those who have taken receipts for their monies as Military Adventurers, be directed to bring in the same forthwith, either to the treasurer, clerk or any of the standing committee of said company, who shall, on their producing such receipt or receipts, or making it evident by any other means, have their names enrolled on the rolls of said company, and be entitled to their part of the land granted and obtained as aforesaid. And those officers and provincials who have served in the late war, who desire to become proprietors and partake of the benefit of said grant, are desired to enter their names as convenient, with some of the receivers appointed by the company for that purpose.

And we do hereby warn a general meeting of the company of Military Adventurers, to meet either themselves in person or by their attornies, for the purpose aforesaid, at the court-house in Hartford in the colony of Connecticut, on Wednesday the 18th day of November next, at 10 o'clock forenoon. And the treasurer of said company is hereby directed, as soon as may be, to have this advertisement inserted in five of the public newspapers of this and neighbouring colonies; and the treasurer is also directed, that at or before the aforesaid meeting of said company he purchase a suitable book or books, in order to record the votes etc. of the said company; and also to enter the names of each one of the said Military Adventurers. Dated at Hartford, the 16th day of October, A.D. 1772

Alexander Wolcott	Hugh Ledlie
David Baldwin	Israel Putnam
Phineas Lyman	Jonathan Waldsworth
David Bull	Joseph Church Jun[ior]
Robert Durkee	Ralph Pomeroy
James Church	Roger Enos[44]
Daniel Bull	

Amid the optimism engendered by the apparent success of Lyman's mission, there was more than a single note of caution sounded by a writer signing himself "Agricola." He asked several pertinent questions, all calculated to cool enthusiasm for emigration to the Mississippi. Where was the written authority to back Lyman's claim? Others asked if emigrants had to attend public officers at Pensacola to obtain a patent? Would not the Indians contest settlement? What support could the emigrants look to in case of war with Spain, since their sole access to the sea ran through Spanish territory? Would not the Adventurers, in the event of a legal dispute, have as their only recourse to go to the courts at Pensacola, several hundred miles distant? Could the emigrants expect to trade profitably when they were located several hundred miles above New Orleans, which itself was difficult to get to from the Gulf of Mexico? Finally, asked Agricola, "What are the motives that can, in the mind of a considerate farmer, overbalance the unspeakable hardships and dangers to which the new planters of Yasou will necessarily be exposed from a hot climate, in a new country, without a fort, and without a port, in the vicinity of jealous Spaniards, and surrounded by the savage barbarians of the wilderness?"[45]

Agricola's skeptical questions were valid. He totally failed to understand (surprisingly, in view of his pseudonym, which indicates that he was probably a farmer) the strong lure of rich, abundant, and free land for New Englanders. This appeal was the main point of an anonymous respondent to Agricola. The land made available in the north since the peace of 1763, particularly in New York, had been snapped up by land jobbers. The poor New England farmers could not afford the prices of these "deceiving monopolizers" and were thus tempted, in spite of hardships, to migrate to West Florida, where, so they believed, there were no land jobbers (a false assumption, although land speculation was still only in its infancy there).[46] The identities of Agricola and his putative refuter remain unknown. One may guess that the latter was an officer of the Adventurers. Agricola may have been a member of a rival land company.

Somehow or other the Company of Military Adventurers had survived for nine years, and on 18 November 1772 a large assembly met in the Hartford courthouse. For three days, no doubt amid much excited speculation, the framework for action by a revived and expanded organization was constructed, although many of the resolutions then passed and the people appointed for service were heard of no more.

The first item of business was to elect a leader. General Lyman was the obvious choice, and he was given the title of moderator. The clerk, or secretary, as we should say today, was Joseph Church, Junior. To explore the Mississippi lands a committee was appointed, comprising Captain Roger Enos, Major John Durkee, Captain John Wadkins, Lieutenant James Smith, Lieutenant Samuel Hawkins, Captain Thomas Hobby, Colonel Israel Putnam, Lieutenant Rufus Putnam, and General Lyman's son Thaddeus. Initially the company meant to send two exploring parties to the Mississippi, one by sea and the other, led by the first four named above, overland and down the Ohio River. The seagoing party, which was the only exploring group ultimately to materialize, was to have a vessel and navigator provided by the company. Each explorer was to be paid eight shillings a day and to receive two months' pay in advance. The explorers were to have assistance from four "able and hardy young men," each of whom was to receive forty shillings a month in pay and thirty shillings a month for expenses. These assistants would be advanced one month's pay.

Since considerable sums were required, a drive for new members of the Adventurers was mounted with a graduated scale of entrance fees: an army officer of field rank had to find ten pounds, a captain six pounds, a subaltern four pounds, a sergeant four dollars, and a private three dollars. Common citizens who were not veterans were now also permitted to become proprietors in the company on payment of six dollars, but those of higher social rank who had no military experience were expected to pay more. The exact sum was left vague, but it was stipulated that it should be more than that paid by former officers.

A new roster of forty-four receivers of dues was compiled. Most of them were located in Connecticut, but some lived as far away as New York and Philadelphia. (See Appendix 3.) Those formerly entrusted with recruiting new Adventurers were forbidden to accept any more, which raises the possibility that abuses had been discovered, but the previous treasurer of the company, Captain Hugh Ledlie of Hartford, continued in his job. Publicity for these decisions was ordered in six newspapers in various colonies.[47] The expense involved may have proved more than initially expected, in spite of the abandonment of the overland exploring party, because, in April 1773, a call went out for receivers to send in the money collected from new members.

The party to reconnoiter the Mississippi was smaller and slightly different from that originally planned. It included Colonel Israel Putnam, Roger Enos, Thaddeus Lyman, Rufus Putnam (a thirty-five-year-old cousin of the colonel), and Daniel Putnam, Israel's thirteen-year-old son. The Hartford meeting had ambitiously asked this committee to try to sail by 10 December 1772. It came admirably close. On 19 December, only a month after appointment, it embarked on the sloop *Mississippi*, which was captained by Wait Goodrich of Wethersfield and manned by a mate, Samuel Foster, and four crewmen.

The *Mississippi* was owned by the firm of Dennis and Smith of Connecticut. She was large enough for trade and equipped for combat, but her main function was to take the small exploring committee to and from West Florida. A prime sailer of sixty-five tons, she had new cables and sails and mounted four brass cohorns, four swivel guns, and four more conventional cannon. Her stern and sides were newly painted green, her hull and rails black. The name *Mississippi* was in gold between the cabin windows.[48] She sailed from New York at noon on Sunday, 10 January 1773.[49]

Evidently conscious of the need for good publicity, the committee advertised its progress. On January 31 Israel Putnam wrote a report from Cape Nichola Mole in the French colony on Hispaniola for insertion in the *New London Gazette,* whence it was copied by other papers. The *Mississippi* had reached Turks Island in eighteen days from New York. She then sailed southwest to Hispaniola in thirty-six hours, where the *Mississippi's* cargo of flour, hogs, sheep, and onions was sold. All aboard were in high spirits, no doubt all the better for having disposed of the livestock accompanying them.[50] They were luckier than perhaps they realized. Three months later the French Chamber of Commerce clamped down on New England trade with Hispaniola, and a frigate, one of two especially commissioned for the purpose, detained several vessels from Rhode Island at Cape Nichola Mole. They were all soon released, but at the same time, having shown their determination and ability to enforce trade regulations, the French issued new ones. Thenceforth only lumber and cattle might be exported to French Hispaniola.[51]

The next stop for the *Mississippi* was Montego Bay, Jamaica, where she lingered only a day. Finally, at sunset on 28 February, the explorers saw the blockhouse on Santa Rosa Island, signifying that at last they had reached the shores of West Florida.

During ten years of British rule, the capital, Pensacola, had developed from the bleak collection of huts bequeathed by the Spanish into a pretty little town with perhaps 200 houses, wharves, warehouses, and a gridiron of streets surrounding a central square, in which stood a fort. In some respects it remained undeveloped, and Colonel Putnam was shocked that Pensacola had no church. By contrast his cousin, perhaps easier to please, admired the "wonderful" gardens with their peach trees and orange trees and greenstuffs of all sorts.[52]

He was also impressed by the hospitality of the merchant James Jones, who introduced the committee members to Governor Chester and at once invited them to dinner, where they discovered the tastiness of Florida oranges. It was the first of several meals with Jones. The other luminaries of the town also made them welcome. The colony's secretary, Philip Livingston, had the senior Putnams to dinner, put them up overnight, and provided breakfast in the morning.

The learned Doctor John Lorimer both dined them and gave them the benefit of his extensive knowledge of West Florida's geography. John Stephenson, the royal contractor, had the committee come to his house to drink tea. Later in that year, 1773, taking tea would be held unpatriotic in New England, but at the time it was not, nor ever would be, condemned in loyal West Florida. Americans in West Florida were not deaf to the call of revolution, but strategic and financial vulnerability made them inattentive. Their colony was heavily subsidized by Westminster and was permanently threatened by internal and external enemies.

The Adventurers' committee was also dined by the Swiss Major General Frederick Haldimand, the senior army officer in the colony. In his thickly accented farewell speech to the Pensacolans in the following month, he referred to "this valuable infant colony which I with pleasure see at length emerging from a state of obscurity and possessed of a prospect of becoming populous."[53] There is little doubt that he referred to the advent of the Military Adventurers. The governor, by comparison with others, seemed initially aloof but actually was delighted to see men of substance in earnest about settling land. A long-held grievance of Chester's was that "injudicious and impolitic" grants to absentees by his predecessors had removed from use the best of the waterfront land along the Mississippi. Chester favored townships rather than huge plantations, and townships were precisely what the new arrivals had in mind.[54]

The cordiality of other leading Pensacolans, which offset the governor's apparent coolness, was surely not disinterested. Philip Livingston, for example, stood to make a fortune from the fees which, as secretary, he charged on land grants if the committee was followed by the promised hundreds of New England families. James Jones too, had sizable business interests in New Orleans and stood to gain much by supplying the Adventurers, since they would all go through the Spanish port. The Pensacola community, however, would probably have cooperated with the Adventurers' committee even if it had stood to gain nothing material from doing so. West Florida was an underpopulated frontier colony and the Adventurers were desirable immigrants. It was natural to welcome them, particularly as they could bring home news to the New Yorkers like Livingston and Jones. General opinion in West Florida favored the arrival of all permanent immigrants, as suggested by the letter of 16 March 1773 by an unknown correspondent who hoped that interest in West Florida would not be confined to the Adventurers:

> Prejudice and misrepresentation have greatly retarded the settlement of this province, but now we begin to surmount their ill effects. The land of the Mississippi and the healthfulness of the climate are so perfectly inviting that emigrants from the northern colonies are daily taking up lands there and improving them, so that in a very few years I hope to see in part of West Florida the most opulent settlement of any in North America.[55]

Of course social warmth mattered less than the legal concessions obtainable from the West Florida council, which, two days after the arrival of the first Adventurers, assembled on Wednesday, 3 March, to receive a memorial signed by the five adults on the committee. Their arrival was not a surprise. Governor Chester knew of the company through newspapers and three months before had received a letter from General Lyman.[56] Explaining that the king had given him a mandamus grant of 20,000 acres, Lyman had stated a preference for a square-shaped tract in the northwest corner of West Florida, bounded on the north by the Yazoo River and on the west by the Mississippi. After discussion with his council, Chester decided that the specified land could be reserved for the general, provided only that another recipient of a mandamus grant not appear, before Lyman's arrival, document in hand, asking for the very same acreage.[57]

Having received the committee's memorial on 3 March, the provincial council told the five members that it would consider it and formally reply to it two days later. The memorial deferentially asked for "such information as in your wisdom you shall think meet." It alleged, on the word of General Lyman, that the king had recognized the services of the Adventurers in the Seven Years' War in spite of their provincial status by allocating, in the same area specified in Lyman's previous letter to Chester, a certain tract of unspecified acreage to *several* Adventurers. The committee understood that relevant royal instructions were to be sent to Governor Chester. Its members wanted to know the governor's attitude to their company and the extent of the land he was empowered to grant it and, if possible, to have a copy of the royal instructions concerning it.

On 5 March the committee went again to the council chamber to hear the answer to its memorial. Chester said he had received no instructions from London relating to the Adventurers; he had, however, received instructions about nothing since August of the previous year and expected none until the arrival of the next mail packet boat. Unless he received instructions ordering preferential treatment, he could do nothing for them beyond what any other immigrants could claim.

The committee may have been disappointed but was surely not surprised. Its members had dined with two council members since arrival in Pensacola and had probably talked to more. They must have expected something like the answer they received. What action, they asked the council, would it then advise? "If," came the cautious response, "General Lyman had good ground for the assurances he had made," then the committee should not waste a benign season of the year but should without delay explore the upper reaches of the Mississippi for suitable township sites before sailing back to Pensacola. By the time of its return, the expected instructions should have arrived. If not, the Adventurers, being "a most respectable and valuable body of people," deserved encouragement in their "laudable plan of settling so fertile a soil." Chester would make sure that they had lands on the most favorable terms possible. At the very least the governor's reply suggested that, in addition to what could be claimed on family right, he would allow an equivalent amount to be acquired by purchase. He ordered the council clerk to furnish them with a certified copy of the royal instructions on granting land; they probably needed no reminder of their terms, but possession of the document would add legitimacy to their mission.[58] Then Chester invited the committee to dine with him and his lady. The governor had probably deliberately avoided close contact with the committee while its petition was *sub judice,* but now he provided a "most genteel and elegant entertainment."[59]

Having once relaxed, the governor became helpful. Pending the arrival of a royal mandamus, he was ready to reserve for two years any tracts that the committee cared to choose, provided only that they were not on land already claimed. Townships of 20,000 acres should be planned. The New York press reported that the committee had been encouraged to earmark five such townships, but actually the number was probably left deliberately vague.[60] Much depended on what was found during exploration, but Chester's attitude was not niggardly: "As these gentlemen appeared to be authorized by a very considerable number of respectable colonists who could not fail of becoming a great acquisition to us I gave them every encouragement that I thought would be consistent with my duty."[61] Unfortunately, in reporting the committee's activities, the press tended not to stress the governor's benevolence as much as the committee's disappointment that he had received nothing relating to them from London.[62]

A week after Chester's dinner party, the committee members were summoned to Livingston's house. The governor, his surveyor general, Elias Durnford, the attorney general, Edmund Wegg, and Livingston told them all that the fees which each of them, in his official capacity, customarily charged for land grants would be reduced. For every 1,000 acres the governor's, secretary's, and surveyor general's fee would be three pounds each, and the attorney general's nine shillings and fourpence.[63] Israel Putnam was not mollified. The fees were still exorbitant.[64] On the same occasion Rufus Putnam was sworn in as an official surveyor, a natural appointment, since he was experienced in the profession; none could guess that one day he would be the surveyor general of the yet unborn United States of America.

Three days later the committee placed its letters home on Alexander Offutt's brig *Betsy and Lydia,* which at this time shuttled regularly between Pensacola and New York. On 16 March the *Mississippi* left Pensacola to begin the long trip that lay ahead. The committee's destination was almost as far from the colony's capital as it was possible to be and still remain in West Florida. The Adventurers had many hundreds of miles to travel before reaching their promised land, much of it involving tedious warping around the bends of the Mississippi, for which, with foresight, the committee had equipped their vessel with a capstan in New York.[65] All the same, because the river below New Orleans was particularly tortuous, the *Mississippi* took two weeks to voyage from Pensacola to the Spanish port.[66]

On the way the Adventurers obtained conflicting impressions of Spanish Louisiana. At Balize, at the entrance to the Mississippi River, they met a schooner loaded with forty French emigrants who could no longer tolerate Spanish rule and were going to Haiti. On the other hand, farther up the river French planters enjoyed impressive prosperity, thanks to the rich soil of the lower Mississippi, where indigo, rice, and corn of surprising height were growing. To Rufus Putnam, the cattle grazing in the fields resembled pampered northern cattle which had fed on hay throughout the winter.[67] The committee also saw several English "floating warehouses" anchored in the Mississippi three miles below New Orleans, whose owners were enjoying the trade advantage conferred on them by the 1763 peace treaty.[68]

No doubt the Adventurers had looked forward to visiting New Orleans. It was small by modern standards, but in the eighteenth century it was by far the largest city of the region, perhaps four times the size of Pensacola. It had been a provincial capital for half a century. Its founder, Jean Baptiste Le Moyne, Sieur de Bienville, had planned it for convenience of trading in time of peace and for ease of defense in time of war. It faced the river. High wooden ramparts reinforced by five forts protected its flanks and rear. Within the fortifications narrow streets crisscrossed, dividing the city into smallish blocks. Eleven such blocks faced the Mississippi, and there were six similar blocks between it and the northernmost city boundary. They contained 700–800 houses, mostly of *colombage,* a combination of bricks and planks. The flavor of the town was French. Spanish soldiers and officials formed a small minority of its population.[69]

Wait Goodrich refused to tie up at the port. The committee members tried to persuade him, pointing out that, since trading was not his intention, there was no possible reason for the Spanish authorities to interfere with him. Goodrich was obdurate and would not cast anchor until he was four miles above the town. Naturally the committee did not stay on board. Quite apart from diversions he may have found there, Colonel Putnam was called to New Orleans by duty. He had been entrusted with a letter from Governor Chester to Governor Luis de Unzaga, who received him coolly. Unzaga was alarmed by the vulnerability of his colony to British invasion if an Anglo-Spanish war should break out, which, following a dispute over the Falkland Islands, was a much-canvassed possibility in the early 1770s. That the population of the Mississippi shore opposite his colony might soon be strengthened by hundreds of New Englanders was not good news, and he did nothing to encourage their vanguard.

In other ways too New Orleans proved a disappointment. Smallpox raged, a serious fire broke out, and the mate of the *Mississippi* broke his leg and had to be left in the town hospital.[70] Yet another disappointment came when Goodrich informed a surprised committee that he would take his vessel no farther up the river. The members had wanted to go as far as Manchac, a hundred miles beyond New Orleans, which they had been told was possible without warping. Wait was probably justified in refusing.

In the days before the steamboat, a variety of vessels could be found for voyages up the Mississippi, from small skiffs made from a single tree trunk to aquatic equivalents of Rolls-Royces, in which the richest planters would travel to and from New Orleans and show off their wealth. A couple of dozen slaves would ply multicolored oars while their master lolled in a pavilion in the stern. The cargo bateaux were sometimes as big as these luxury craft. Over their raised poops were covers of canvas and oxhide. Midships the cargo was protected by a tarpaulin, on each side of which sat rowers. In the bow sat the boatswain, constantly taking soundings, a necessity on the treacherous Mississippi. All these river vessels, whether flat bottomed or keeled, were propelled by oars. Depending on the state of the bank and river depth, warping was sometimes an option, but sailing was a less practicable alternative to rowing; along the course of the twisting Mississippi in many places tall trees on either shore cut off the wind from the water.

The whole art of swift navigation upriver consisted in knowing and using the eddies in the Mississippi and avoiding the sandbars, floating debris, and places where the current or offshore winds were strong. Under even the most experienced captain, a fully loaded forty-ton bateau would go at best eighteen miles a day up the Mississippi.[71] Such a vessel normally took a week to voyage from New Orleans to Manchac. The *Mississippi* would have taken a fortnight, even if Goodrich did not damage her by running her aground, which was all the more likely in that she was not flat bottomed and, as a seagoing sloop, probably of deeper draft than most river craft.

In all, going to Manchac would have lengthened Goodrich's voyage by at least three weeks. Goodrich was probably wise too in refusing to anchor off New Orleans, since only a few months previously Unzaga's men had taken and auctioned the Rhode Island vessel *Two Pollies* on a trumped-up charge.

Newspaper readers interested in the Adventurers were misinformed that their committee left New Orleans on 18 May.[72] In fact it had continued upriver over a month previously in a barge belonging to Elias Durnford, the lieutenant governor of West Florida.[73] By 15 April its members had ascended as high as the confluence of the Mississippi and the Iberville, about sixteen miles below modern Baton Rouge. The Iberville, which could be seen on any map to connect the Mississippi to the Gulf of Mexico by way of Lakes Maurepas and Pontchartrain, throughout the period of British rule inspired projects designed to funnel trade from the upper Mississippi and the Illinois country through this water link. The Adventurers could bear witness to the fact which thwarted all such dreams: the Iberville was seldom navigable. When they saw it, the Iberville was so low that thousands of evil-smelling fish lay on its bed, along which the committee members walked with a British engineer whom they chanced upon, a Lieutenant Hutchins. Naturally he did not tell them so, but Hutchins was engaged in spying. General Thomas Gage had employed him in that year to plumb Spanish military strength in Louisiana and to sketch the fortifications of New Orleans.[74] Perhaps filled with enthusiasm by the Adventurers' conversation, on 2 June 1773 Hutchins petitioned for 25,000 acres south of Natchez, with the intention of settling there immigrants from the middle colonies, a scheme which came to nothing.[75]

Soon the Adventurers were beyond Manchac, a small collection of "tolerable" houses with gardens, only one of which, perhaps John Fitzpatrick's, impressed Rufus Putnam. In succeeding days the committee barge passed nothing more interesting than scattered English plantations on the Mississippi's east bank until arrival at Pointe Coupée on 19 April. This sprawling French community on the western shores stretched an incredible twenty-four miles, along which houses were regularly spaced at gunshot distance from one another. Although the land tracts to which they belonged had narrow frontage on the river, they extended far back from the bank. After the committee sailed past the old Tonica village on the English side, the river became particularly tortuous, and the rate of its progress slowed to about sixteen miles a day, whereas lower down the river it had traveled as much as forty miles in one day. Finally, on 26 April, the Adventurers reached Natchez, the first significant British settlement since Manchac. It was a focal point for inhabitants living in its vicinity rather than a town with shops and streets. Its main building was Fort Panmure, which could house a garrison of fifty, although no troops had been stationed there since 1768. Apart from the disappearance of the barracks, which had been burnt down by Indians during an alcoholic spree, the fort was in good condition. There were even cannon still in place.

The Adventurers' committee soon found that it was far from alone in scouting out desirable land on the Mississippi. For some time there had been an influx from the backcountry of Virginia, Pennsylvania, and the Carolinas. Refugees from and participators in the Regulator risings were among those who sought a fresh life by traveling overland and squatting on the rich lands of West Florida on or near the big river. Many of them were indigent. Some of them were criminals. Governor Chester did not discourage any immigrants, although he did describe such settlers as "first immigrants" and unsuitable for responsibility. When a considerable number of poor migrants from the backcountry of other colonies settled in the Natchez district, Chester did not turn them away. He even lightened their hardship with expenditures from his contingency fund, which he was aware was not intended for any such purpose.[76] Undoubtedly he welcomed the Adventurers even more warmly because they were potentially fit for responsibility. He could visualize them as justices of the peace or representatives in the West Florida assembly.

At Petit Gulf, north of Natchez, the Adventurers met Major Luke Collins, who had brought a party of twelve migrants with him from Redstone, Pennsylvania, on the Monongahela, about thirty miles south of Fort Pitt. Two days later, on 3 May, they met another squatter from the Pennsylvania backcountry. A rush was on, and the Military Adventurers was not the only organization favored by Chester. In 1770 eighty pioneers had arrived at Natchez from Redstone. In 1771, 200 families under Colonel John Clark migrated there from the Holston River, a tributary of the Tennessee, and William Canty brought a party of thirty settlers to the Mississippi.[77] In 1772 Captain Ogden, as has been seen, brought in settlers from New Jersey.[78]

Even while the Adventurers' committee was scouting the Natchez area, Samuel Sweezy was petitioning for lands there for a group of families and slaves numbering sixty-eight which he had already brought into West Florida. On the same day, 19 April 1773, Jacques Rapalje, on behalf of his father, Garrett Rapalje, a New York merchant, asked for 25,000 acres to be reserved on the Mississippi for a group of settlers he intended to bring into the colony. He asked for, and received, the same preferential terms which, so he must have heard, had been granted to the Military Adventurers.[79]

Immigrants from English colonies did not necessarily choose to settle on the English side of the Mississippi but sometimes preferred the better-developed Spanish shore. A correspondent at Pointe Coupée on 23 June 1773 reported that 120 families had come down from the Virginia and North Carolina backcountry to settle there. The prospect of an increase in the population of the potential enemy may well have persuaded Governor Chester to be more indulgent to entrepreneurs than otherwise he might.[80]

The voyage in July of a brig from Elizabethtown, New Jersey, under Captain Thomas Davison, was probably part of an entrepreneur's scheme. The brig carried seventy passengers intending to settle on the Mississippi. In the same summer twenty-nine boats full of settlers, possibly an ad hoc conglomeration rather than an organized party, arrived at Natchez from the north.[81]

They undoubtedly had competitors, but the Adventurers, amid the vast expanse of province, must have feared overpopulation less than other settlers' reservation of the choicest tracts. They pressed on fast, looking over a very large area in a comparatively short space of time, in the region north of Natchez.

Almost every day in the beginning of May, the Adventurers met Indian traders and new settlers. Jacob Winfree, for example, had arrived from Virginia with an unofficial group of 100 people, which included women and fifteen blacks. Winfree intended to establish a plantation and indeed the following month would be granted 1,000 acres on Second Creek.[82] Potentially more likely to thwart the Adventurers' plans were the local Indians. On 5 June they met a Choctaw who swore that his tribe would not allow anyone to settle on land above the Big Black, because it was the only hunting ground left to it. This conviction was no isolated opinion. Nine days later they asked a Choctaw why he refused to accompany them on their trip and show them the land. He told them that he had met two of his chiefs who rejected the Adventurers' casuistic tale that they were there to see the land not to settle it. The chiefs believed, accurately enough, that, if the committee liked what it saw, then white settlers would follow. Although the Choctaw could grow their own food, they needed to hunt too, because only through trading skins and furs could they acquire the blankets and clothing that they and their families required. They had been told by Superintendent Stuart that there would be no settlement above the Big Black. If any such attempt were made, not only the Choctaw but the Chickasaw too would resist, even though both tribes liked the English.[83]

The members of the committee seemed quite undeterred by the possibility of Indian hostility, even splitting up into small groups to explore more thoroughly the region between the Big Black and the Yazoo, where they intended to establish townships. Sixteen or seventeen miles up the Big Black, or Loosa Chitto as it was also called at the time, Rufus Putnam and Thaddeus Lyman found suitable land for a town which was "reserved for the capital." It seems that the endless speculation in the newspapers of the day that a new colony would be created was not merely a possibility for the Adventurers; it was accepted without question.[84]

Having traveled as far as the junction of the Yazoo and the Mississippi, the Adventurers descended the big river in the last days of May. They were able to rejoin their vessel *Mississippi* on 3 June, about twenty miles below Manchac. They continued to find proof that 1773 was a boom year for immigration to Florida. On 14 June the *Mississippi* passed a sloop going upriver. It was from North Carolina and contained thirty-three passengers intending to settle.[85]

Initially wary of visiting New Orleans, the Adventurers entered it on 19 June once they were sure that the smallpox epidemic had ended. They called on the Spanish governor, who gave them permission to roam the town freely and who politely, if not entirely sincerely, assured them that he would do anything he could to help them. The Adventurers' committee was probably glad to rest at New Orleans because one of their number, William Davis, was ill with fever. They stayed ten days, dining with one Hazzard, an English merchant, took on a load of wood for ballast, and finally acquired two passengers. One was Anthony Hutchins, whose brother they had already met at Manchac. The other was Patrick Morgan, of the New Orleans partnership of Morgan and Mather.

By 5 July the *Mississippi* was anchored off Pensacola. In many ways the committee's reconnaissance had turned out fortunately. Its members had all returned safely. The lands they had viewed had largely lived up to optimistic expectation: they were fertile and well watered, and, but for the Indians, whose rights the committee clearly did not intend to respect, they were empty. The climate and soil were obviously suitable for farming, if the French side of the Mississippi River was any indication. The Adventurers had explored their future homes in the spring, when the weather on the Mississippi is not oppressive, and so probably felt that the climate favored people too. To add to their general satisfaction, they had enjoyed fine hunting, shooting buck, bear, and turkey, in addition to catching catfish of extraordinary size.[86]

Of more consequence they had, working with admirable expedition, tentatively seen and earmarked for townships nineteen different sites. One of the first things they did on arrival at Pensacola was to engage the provincial surveyor general to make a map of the area they had visited and, from their descriptions, bearings, and rough sketches, to locate on it the potential townships so that it would be possible for the council and others to see precisely where the lands lay that they wanted to reserve.

One thing only was disappointing. In their absence no instructions had arrived at Pensacola from London relating to the 150,000-acre tract for the Company of Military Adventurers. The main difference it would make, if and when it did arrive, was economic. It would mean that seven townships could be completely settled before a penny of purchase money had to be found. Otherwise they were "perfectly satisfied."[87]

On the same day, 7 July, that Elias Durnford was drawing maps for the committee, the West Florida council met and considered the request of the Lyman family. Thaddeus was agent for his father, who had stayed in New England but had entrusted his personal mandamus grant for 20,000 acres to his son; having looked over the countryside, Thaddeus had decided to locate their land on the Bayou Pierre. He intended to place a township there called New Suffield and asked for issuance of a warrant of survey. The council consented.[88]

Two days later, on 9 July, the Adventurers again met the council, described their recent trip, and asked that nineteen townships of about 23,000 acres each should be reserved for them for a year. The council promised to reserve them only until 1 March 1774 but pronounced its willingness to extend the time limit if the Adventurers produced proof that their families were preparing to emigrate. When sufficient families arrived to qualify, under the family right regulations, for one-half of the acreage of a township, then the other half, promised the council, would be made available for them for purchase. The average Adventurer family numbered seven, and each member would qualify for 50 acres—except for the family head, who was entitled to 100 acres—which meant that this proviso could come into effect when two dozen or so families decided to settle in a given area.

The committee men seemed confident that not mere dozens but hundreds of families would want to come to Florida. They vied with each other in securing the right to people particular townships. Israel Putnam wanted number nineteen on the plan which had been circulated among the councillors, Rufus number six, and Enos number five, while Thaddeus Lyman, not content with helping his father settle New Suffield, asked permission for himself and his associates to settle township number seventeen. Actually the entire committee, with the exception of the teenager Daniel Putnam and the ailing William Davis, wanted the responsibility of settling townships. Even Wait Goodrich, a sea captain by profession, was interested enough to "undertake," as the phrase then went, township number eighteen. The council replied cautiously that they were prepared to reserve the townships only for the company, not for individuals, unless the company were to dissolve.[89]

A few days later the explorers left Pensacola. West Florida's provincial secretary, Philip Livingston, remained to handle company affairs. He probably had an interest in it prior to the arrival of the exploring committee, because his father, Peter Van Brugh Livingston, was an official representative of the company in New York. The younger Livingston told the provincial council on 2 August that he had heard from the exploring committee that the Adventurers' standing committee in Hartford had confirmed that settlement in West Florida would proceed, even if no royal mandamus arrived authorizing the longed-for large land grant. Livingston therefore asked that the time limit on the reservation of the company's projected nineteen townships be extended to eighteen months. When the council demurred the secretary modified his request. Would the board then reserve a mere four townships for eighteen months? Again the council refused. Perhaps they doubted that, in the short time that had elapsed since the explorers left, the standing committee would have had time to receive a full report from them. In fact the standing committee had received no report of any sort because the *Mississippi* was still at sea, not anchoring in New York harbor until 7 August, when Enos, Lyman, and the Putnams disembarked and headed for New England.[90]

The Adventurers' standing committee did not communicate with West Florida for several months, and then, quite properly, it sent a letter to Governor Peter Chester. The news was encouraging. A full meeting of the company had been held in Hartford on 5 November. Between 400 and 500 members had attended, most of them the heads of large families. The vote in favor of settling in West Florida without delay was unanimous. Four sets of undertakers, or contractors, had been appointed to settle four townships, and they were instructed to sail by the tenth of December.

Enthusiasm seemed undimmed—which was surprising, since confirmation of the land grant of which Lyman had boasted had never come; on the contrary, it was now well known that on 7 April the king had ordered an end to all land grants except for the limited sort specified in the 1763 proclamation.[91] It was also known that General Gage, who opposed new colonies in the interior as certain to provoke war with the Indians, had sailed for England in the summer of 1773. Later it was confirmed to Americans that Gage had made this view known to the ministry in London.[92]

At the same time it was plain that Governor Chester was cooperating as far as he could with the Adventurers, while their exploring committee had proved that there was indeed rich land available in his province. An added inducement was the impression conveyed by the newspapers that other settlers coveted the land. In the summer of 1773, according to one press report, 400 families from the backcountry of Virginia and the Carolinas made their way down the Ohio to settle in West Florida. It was also reported, probably quite falsely, that four emigrant ships from North Carolina crammed with settlers reached West Florida by sea.[93]

And so, almost on schedule, two Adventurer ships left Connecticut in the latter part of December 1773. The first was the old *Mississippi,* commanded this time by Captain Porter of Stonington. She left New Haven on 17 December but, delayed by bureaucratic formality, had to voyage to New London for a customs clearance before starting properly on the last day of the year. She carried about fifty passengers, of whom at least thirty-six, according to a list compiled by one of the Adventurers, Matthew Phelps, were male settlers.[94] One of them was General Lyman's third son, Thaddeus. With him Thaddeus took eight slaves and his semicapable elder brother Phineas.[95] This vessel made good time and en voyage caught up with the other company sloop, which was suitably named *Adventurer* and was commanded by the *Mississippi*'s former captain, Wait Goodrich. She had sailed from Middletown on 19 December,[96] and had touched at New York, which she left on 13 January 1774.[97] *Adventurer* also contained fifty passengers, including the old general and other members of the Lyman family,[98] as well as Timothy Ledlie, son of a Hartford merchant who had recently lent the general in money or in kind over £1,000 (Connecticut money), which he would never see again.[99] Quarrels, however, lay in the future. For the time being there was harmony between the Ledlies and the Lymans aboard *Adventurer.*

It may seem surprising that Israel Putnam, the senior member of the exploring committee and one of the most ardent in seeking to establish a township, subsequently should not have been on one of the first immigrant vessels. He apparently was distracted from his participation in the Company of Military Adventurers by involvement in prerevolutionary political activities at home. As early as 28 June 1774, he was chairman of a standing committee of correspondence for the parish of Brooklyn in Pomfret, Connecticut. His main business was initially, in sympathy with Bostonian resistance to Britain, to prevent the importation and consumption of any goods from Britain and India.[100] It was of course a mere stepping-stone to a more important role in the revolution.

It is difficult to assess accurately the extent to which the worsening dispute with Britain cooled enthusiasm for emigration, but it may have been considerable. Settling within the boundaries of British West Florida meant trusting in the benevolence of the British king, ministers, and Parliament to an unusual degree, at a time when colonial opinion was attributing the most selfish of motives to Britain's every act. By contrast it was thought that, if the colonies should achieve some kind of autonomy—and both full and partial independence were discussed long before 1776—then the British restrictions on settlement beyond the Alleghenies which had existed since 1763 were sure to be abolished. In that case there would be no need to voyage to the distant and foreign Mississippi to provide for one's children. Good land would be available closer to home.

When the *Adventurer* sloop docked in Pensacola on 3 March 1774, all aboard were in perfect health and high spirits. Their coming presented Chester with a dilemma concerning bounty lands. The governor had been cautious and niggardly in granting lands in accordance with the 1763 proclamation. He insisted that claimants produce certificates of wartime service unless, himself a war veteran, he knew the applicant's claim to be valid from previous acquaintance. Moreover, for long he entertained no applications, however valid, from absent petitioners. He relaxed this restriction only on the orders of the plantations secretary. Lastly, from the time of his becoming governor, he had refused to entertain claims from veterans who had served in provincial units in the Seven Years' War.

There would have been no need for Chester to contemplate reversing this policy if the huge mandamus grant confidently expected by the Adventurers' exploring committee had ever arrived in West Florida, but it had not. To make his dilemma more acute, on 7 April 1773 George III had forbidden all land grants on purchase or family right until further notice. The only exceptions to the royal moratorium were grants to veterans qualifying for land under the terms of the 1763 proclamation. If, for the first time, Chester allowed lands to provincial veterans, then he could expect the Military Adventurers, highly desirable immigrants, to stay in West Florida. If he did not, they might return to New England or look elsewhere for settlement, in Spanish Louisiana perhaps. But reversing his policy toward provincial veterans would lay him open to the charge of inconsistency and favoritism. Initially he made no decision on the subject but sensibly waited to hear what the Adventurers had to say.

Their spokesmen, Major Timothy Hierlihy, Captain John Elsworth, Hugh White Junior, and John Kirby, presented a memorial to the West Florida Council. It was signed too by Thomas Lyman of Durham but oddly, since he was also a passenger on the same vessel, not by General Phineas Lyman. He was certainly in Pensacola at the time. Two days after his arrival there, he had struck a bargain with three leading citizens, Elihu Hall Bay, Alexander McCullagh, and Philip Livingston. In exchange for 3,000 acres carved from Lyman's 20,000 mandamus land, the trio agreed to pay the general £300 in New York currency.[101] The memorialists accurately recounted the activities of the exploring committee and of the senior Lyman. They knew of the royal order of 7 April 1773 prohibiting further land grants but hoped that the encouragement given to the Adventurers' committee prior to the arrival of the king's prohibition would justify their occupying four of the townships earmarked for them in the northwest corner of the province. On the basis of this encouragement, they alleged, 104 emigrants, each with a family averaging seven members, were either in, or on their way to, Florida.

The council was sympathetic but cautious. It said that Hierlihy's memorial should go to London for a decision by the plantations secretary. In the meantime the nineteen township areas in which the Military Adventurers planned to settle would be reserved for them for six months.[102] This decision was prudent rather than discouraging. Certainly there was no reason for the immigrants to turn back. They could continue to the proposed townships and squat there: competition from other potential settlers in that remotest of all corners of the province was unlikely to be serious. If Lord Dartmouth approved the townships, they were off to a flying start; if not, they could squat and expect to obtain the land free under family right or by purchase once the royal ban—which was rightly assumed to be temporary—was rescinded. Either way the Adventurers should stay, particularly as Chester was eventually persuaded that he should honor provincial service as satisfactory qualification for proclamation land.

Chester admitted that he did so at the urging of several Pensacolans but chiefly because of the arguments of Hierlihy and the Lymans. All three of the latter had served in Cuba in 1762 and had certificates to prove that, just like regulars, they had been considered equal with them when it was a question of allocating the prize money for the capture of Havana. Second, they reminded Chester that previous governors of West Florida had given bounty land to provincial veterans. Finally, they cited the practice of other colonies. In Virginia and New York provincial veterans had been granted proclamation land.[103]

Thus encouraged, the Adventurers continued on to New Orleans, where they disembarked and made arrangements to proceed farther upriver in smaller craft. The area where the Military Adventurers intended to establish themselves lay on the Big Black River or, where General Lyman himself would settle, on the Bayou Pierre. With Phineas and Thaddeus he arrived in the spring, intending, without question, to be a leader of the infant settlement which he had done so much to initiate. On 5 March the West Florida council voted, doubtless with his acquiescence, that he was a "very fit and proper person to be a magistrate on the Mississippi."[104]

And so General Lyman probably would have been. But the fifty-nine-year-old veteran, though seasoned in war, was not able to fight the climate in the sickly summer season and was attacked by one of the now unidentifiable fevers which carried off many of the Military Adventurers. Forced to consider the possibility of abandoning his own hopes for a life in West Florida, the patriarch still foresaw a life there for his large family. On 16 August 1774 he petitioned the governor that, if he died, his new plantation, Nanachay, should be divided among his children, with Thaddeus obtaining the lion's share. He also specified another division among the remainder of his family if Thaddeus, who was also ill, should die.[105]

A word of explanation is probably needed as to why Thaddeus, the third son, rather than his older brothers, should have received preferential treatment. The senior son, Phineas Junior, was born in Suffield on 21 September 1743. As a spirited Yale student he had headed a class rebellion against tutors, for which he was rusticated but restored to the rolls in time for graduation. Thanks to his father's influence he was then commissioned in the British army, which he abandoned to study law. He qualified as a lawyer but practiced waveringly.[106] His mental stability deteriorated. Finding him in sad condition after his sojourn in England, General Lyman sent him with Thaddeus to the Mississippi on 17 December 1773 in the hope that transfer to a new climate would mend his health and spirits.[107] It was not to be. When disposing of his property his father wrote matter-of-factly that "his eldest son" (he seems to have avoided using his name deliberately) was "habitually delirious and must be supported by Thaddeus if he lives."[108]

The second son, Gamaliel Dwight Lyman, was similarly excluded from the general's Mississippi plans. Eighteen months younger than the junior Phineas, he was "a man by nature gay and ingenious." It was he who had been chosen to go to England and to persuade his father to return. While there he had been given a lieutenancy in the sixty-sixth of Foot. His father gave two reasons why it was unnecessary for Gamaliel to have land: not only was he already "genteelly" provided for in the British army, but he also had "no disposition for agriculture."[109]

But even if Thaddeus died and Gamaliel were absent, there were still other children to look after young Phineas and to claim a share in Nanachay: two sons, Oliver and Thompson, and two daughters, Eleanor and Experience. As it turned out, perhaps luckily, not Thaddeus but the delirious Phineas Junior died, evidently shortly before his father did on 10 September 1774.[110]

Undeterred by the death and sickness that had visited the first three Lymans to settle on the Bayou Pierre, the rest of the family, Gamaliel excepted, left Middletown for the Mississippi on 1 May 1776, together with many other Military Adventurer families. After a voyage made hazardous by war, they arrived at New Orleans on 18 August.

There they transferred themselves and their chattels to canoes and bateaux and proceeded upstream to Bayou Pierre, where Thaddeus alone remained to greet them at Nanachay. It is sad to relate that the brave matriarch, Mrs. Eleanor Lyman, followed her husband to the grave within a few months of her arrival there, but the rest of her family stayed on.

Thaddeus, meanwhile, as head of the clan, had succeeded his father as justice of the peace for the Natchez district on 2 January 1775.[111] He took steps to have his father's 20,000-acre tract transferred to his own name, promising that he would then transfer parts of it to his brothers and sisters in fee simple, in accordance with his father's wishes. Governor Chester was sympathetic to the request, which was made on 28 January 1775, and approved the transfer legally on 2 February. The survey of the tract was recorded on the same day.[112] Twelve days later Thaddeus sold 1,000 acres in the northwest corner of this vast tract to John Brown, the barrack master at Pensacola, for 1,000 Spanish dollars.[113]

Although all the legal formalities were apparently complete, thereafter something went wrong. For some reason the grant must have been found invalid, for the next we learn of Nanachay is that members of the Lymans were applying for it on family right—a tedious process involving physical appearance before a magistrate of all on whose behalf claims for acreage were being made. This tiresome procedure would surely have been avoided if it had not been absolutely necessary, especially as it would yield the Lymans less land than provided by the original mandamus grant. However that might be, on 13 November 1778 all of the general's surviving progeny in West Florida applied for and received tracts on the land they had already settled. Thaddeus applied for Nanachay itself as the holder of seven slaves, obtaining

450 acres on family right and a further 1,000 acres as loyalist bounty. His younger brother Thompson received 600 adjacent acres, as did his other brother Oliver. Their sisters, Eleanor and Experience, each claiming a black maid to fortify their petitions, both received 650 acres on the Big Black River. So the family emerged with 3,950 acres, which must have been more than enough to meet their immediate needs.[114]

Thaddeus was a responsible young man. At the beginning of 1779 he described himself as representative of the Natchez district in the West Florida legislative assembly, justice of the peace, and "lately a captain of an independent company at the Natchez."[115] Under his leadership the Lymans seemed well established.

Meanwhile West Florida's acting provincial secretary, Philip Livingston, was using his Adventurer connections to further his personal ambitions. Just as Livingston became the company's agent in West Florida, so also, it seems, did Roger Enos of Windsor, Connecticut, the former member of the exploring committee, become Livingston's personal agent in the northern colonies. Livingston enjoyed the confidence of Governor Chester, accumulating an extraordinary number of official positions in West Florida, and was bent, in his private capacity, on maximizing his landholdings. His critics eventually charged that he amassed 150,000 acres, although his biographer has determined that 50,000 would be a closer figure.[116] One of his techniques was to buy up the rights of reduced militia officers who, as war veterans, could have claimed land in West Florida under the terms of the 1763 proclamation but who had no desire to emigrate there. Enos was employed to find old comrades in this position and to conduct the legalities necessary to transfer their rights to bounty land. They had all served under General Lyman in the First Connecticut Regiment during the Seven Years' War. They were Captain Noah Humphrey, Lieutenant Elihu Humphrey, Ensigns Nathaniel Humphrey and Jonathan Pinney, and Captain Zebulon Butler. Each captain could claim 3,000 acres, and the others 2,000. Thus Livingston acquired the right to 12,000 acres. The veterans were not paid according to a fixed scale, and the Humphreys were worse rewarded than Pinney and Butler. Nevertheless Livingston paid them all a total of £330 in New York currency, as well, it may be presumed, as the expenses of Enos, who traveled between his native Windsor and Sunbury, Hartford, and New York in performing his business for the secretary.[117] The five potential settlers must have rejoiced at the bargain they had made when the revolution rendered all their claims to crown land worthless.

It seems likely that by 1775 Enos had abandoned his original intention of settling in West Florida. He did not sell his own claim to bounty land to Livingston. Instead, acting through his attorneys, James Hamilton and James

Amoss of Pensacola, Enos sold the right to his 1,000-acre entitlement to a West Floridian consortium consisting of David Hodge, Samuel Fontenell, William Wilton, and John Crozer for £150. The claim was subsequently presented and honored, and on 13 October 1775 a tract on Bruce's Creek, northeast of and three miles back from the Amite River, was patented in the name of Enos, although he had resigned his entire interest in it.[118] He had received substantially more for his claim per acre than had his old comrades.

Escorting old General Lyman's widow to Florida had been her brother Major Timothy Dwight of Northampton, Massachusetts. He was a man of exceptional physical and moral strength. Six feet four inches in height and of large build, he had once, for a joke, proved that he was literally as strong as an ox. He could support his petite wife, it was said, on the palm of his outstretched hand. His remarkable probity is illustrated by another tale. Princeton College held a lottery. President Burr, Dwight's brother-in-law, sent him twenty tickets to sell. Because Massachusetts passed a law against selling lottery tickets from another colony, Dwight put them aside, intending to return them except for a particular one for himself. The lottery rules specified that all tickets not returned by a given date had to be paid for by the holder. That day passed without Dwight's returning any of the tickets in his possession. The drawing took place, and one of these tickets won the best prize of £4,000. Dwight refused to claim the money because the winning ticket was not the one that he had set aside for himself. Instead he merely sent in payment for all the tickets. His usual conscientiousness was perhaps strengthened by his wife, a daughter of the preacher Jonathan Edwards.

By profession Dwight was a judge on the court of common pleas. As such he had taken an oath of fidelity to the British crown. When revolutionary feeling rose in Massachusetts, Dwight was stripped of his civil and military offices, and a gang surrounded his house and threatened to destroy his family. He emigrated to the Mississippi with his sister and two of his sons, Sereno and Jonathan. His wife and the rest of his children, who included a future president of Yale, he left behind in Northampton. Dwight intended to settle, applying on 6 January 1777 for two tracts in West Florida's northwest corner, but, as strong as he was, the climate of the province proved too much for him, and he died on 10 June 1777, two months after his sister.[119]

Traveling on the same vessel as the unfortunate Major Dwight and Mrs. Lyman had been the Flowers, Leonard, and Smith families, each with numerous children. Mrs. Flowers died at Pointe Coupée in August 1776. A little farther up the river, Mrs. Leonard died at Natchez, as did the Reverend Jedediah Smith two days after his arrival there.[120] As so often happened, the survivors stayed on.

Sereno Dwight, Major Timothy's eldest son, a Massachusetts apothecary, had been imprisoned for toryism. He fled Northampton with his wife and children in April 1776. His money was depleted by the cost of treating his father's fatal illness. Perhaps he could not afford to return. He was granted 1,250 acres on the Bayou Pierre, adjacent to Thaddeus Lyman's, about three miles from the Mississippi.

On the same day that Dwight received his grant, Josiah Flowers of Springfield, Massachusetts, whose wife had died two years before, applied for land next to Nanachay and was given a family right grant as though she had not. Receiving some bounty land as well, he was given 400 acres on the south side of the Bayou Pierre. Abijah Leonard, whose mother had died at Natchez in the same summer as Mrs. Flowers, was not so lucky. He received a single man's grant of 100 acres on the Big Black near the Lyman girls' tract.[121] Again Sarah Smith, a fellow passenger of the Leonards, whose husband, Jedediah, a congregationalist minister, had died prematurely, did not return to Gainesville, Massachusetts, but stayed on, applying successfully for a land grant. The council was uncommonly generous to her. As head of a household she received 100 acres northeast of Natchez, another 100 acres as a bounty for loyalty to the crown, and 1,000 acres for her ten children.[122] Several of them, Joseph, Philetas, Philander, and Calvin, settled near Natchez and established large and influential families. Others, Luther and Courtland, settled near Bayou Sara and also had many descendants.[123]

Voyaging with Mrs. Smith had been Benjamin Day, a New England wool merchant, who had brought ten children and three slaves with his wife Sarah (née Dwight). The council allowed him only 400 family-right acres on the north side of the Big Black River, although he was also given 1,000 bounty acres. Another passenger traveling with Day was yet one more Dwight, Henry, who applied for land on the Big Black two miles from the Day plantation. He received 200 acres, which, as the owner of two slaves, was the minimum to which he was entitled on family right. Despite his professed detestation of the revolution, he received no loyalist bounty land.[124] Finally, this boatload of Adventurers included John Felt of Suffield, who, on behalf of himself and his wife and child, likewise received a minimal 200 acres on the Big Black. Perhaps his loyalty was suspect—and with reason. In 1779 a man with the not particularly common surname Felt joined the West Florida provincial defense force, deserted from the Natchez garrison, and shot to death a Lieutenant Pentecost who tried to recapture him.[125] The officer was probably George Pentecost of South Carolina, who abandoned considerable property in debts to migrate to West Florida in June of 1776.[126]

The vicissitudes of another Military Adventurer, Matthew Phelps, still make fascinating reading after nearly two centuries. Others may have experienced equal joy and horrors, but Phelps was the only one to write a book about them. Required for health reasons to move from Connecticut in 1773, when "there was much talk about the goodness of the country near the Mississippi," he sailed from New Haven in December. In the Dry Tortugas off the southwest tip of Florida an incident occurred that is reminiscent of the fictional *Caine* mutiny: the mate of the sloop on which Phelps was a passenger forcibly dispossessed the captain of his command.

Phelps reached the Big Black River, 120 miles above New Orleans, without further excitement, where he purchased a tract of land from a squatter for fifty dollars. Leaving it in the charge of an overseer, he returned to Connecticut, pleased by the prospect in West Florida and determined to take his wife and family there. In spite of a summer illness which delayed him in New Orleans on his way back and a near-wreck off Cape Hatteras because a chest of blacksmith's tools near the binnacle had skewed the compass, Phelps was able to rejoin his family in August 1774.

The revolution was well under way before Phelps finally sailed for a second time, in May 1776, with the Dwights and Lymans and other Military Adventurers. His journey was doleful from Manchac onward. In September his daughter Abigail and his infant son Atlantic died of a fever. So too, at Petit Gulf, did his wife. Finally, while ascending the Big Black River, his remaining children, Ruth and Luman, were drowned in a whirlpool. Alone he arrived at last at his plantation to find it deserted by his overseer and occupied, legitimately according to the custom of the area, by strangers. With help from a certain John Storrs whom he aided on his first expedition, he reestablished a new plantation and soon prospered.[127] His title to it was subsequently legalized when the provincial council granted him 350 acres on the north side of the Big Black about twenty-seven miles from its mouth.[128]

During the revolution Phelps was a soldier for the British, a deserter, a brewer, a merchant, and a boatbuilder. Dismayed by the Spanish conquest of the area where he lived, he sailed back to Connecticut via Havana in 1780.[129]

More successful in planting his family permanently in the Mississippi area was Daniel Lewis. His father, Daniel Lewis, Senior, came from Sheffield, Massachusetts. The younger Daniel there married a member of the distinguished Fairchild family. Later he migrated with the Military Adventurers to Natchez with his wife, two children, and six brothers and sisters, abandoning land in Connecticut for which he could get no price, or so he alleged, because of revolutionary disturbances. His father and mother had traveled with them but had died soon after reaching West Florida. Lewis was granted 500 acres on the south side of the Bayou Pierre close to the Lyman tract.[130]

He later drowned on a trip to New Orleans, and his widow twice remarried. Her second husband was Richard Carpenter, a Rhode Island Quaker, and her third, General George Matthews of Georgia. She died in Mississippi in 1803. Her first child, Archibald, became prominent in the Mississippi Territory.

For some time the Adventurers could pursue their private interests as though the fire and blood of revolution did not concern them. This temporary tranquility was broken in the winter of 1778 by the return of James Willing, who had been a planter in the Natchez district of British West Florida.[131] With a naval captaincy, a lieutenant called Thomas McIntyre, and a well-armed force in hunting smocks, he came, ostensibly to win West Floridians to the revolutionary cause but actually to loot and destroy. Willing was lucky that his progress downriver was unopposed. The Choctaw sentinels on the Mississippi were carelessly changing the guard at Walnut Hills, and he managed to slip past this dangerous spot unobserved; nobody had thought it likely that rebels would attack West Florida in February. On the 19th the American flag was raised over the old fort at Natchez. The inhabitants in that district, deceived into believing that they were threatened by far more than the hundred or so followers that Willing had acquired, pledged their neutrality; several dozen of them even joined him. Thus reinforced, Willing continued downstream, sacking the English settlements at Baton Rouge and Manchac before Governor Chester at Pensacola had even heard of his arrival.

Willing's success demonstrated a weakness in Britain's policy toward West Florida. The province was so large that its western region could not be held without permanent garrisons in the posts on the Mississippi. As one writer put it, the inhabitants in the west could as well look for protection from the Tower of London as from Pensacola if they were attacked.[132] This assessment was an exaggeration. The Pensacola authorities were capable of an effective, if slow, response to an attack on the Mississippi settlements, particularly, as in this instance, if the attackers alienated rather than conciliated the inhabitants.

Chester relied on two prominent loyalists in the Natchez district to make the response. Neither had taken the oath of neutrality and both were "particularly obnoxious to the rebels." One of them was Anthony Hutchins, who escaped from captivity by Willing and sought to persuade the settlers of Natchez to rise against Willing on the grounds that he had broken his agreement with them. He found little support until he summoned the other subject most trusted by Chester, Thaddeus Lyman, from his remote residence. The arrival of Lyman with friends, most of whom were, no doubt, Military Adventurers, decided the issue. Natchez was recovered for the British. Hutchins and Lyman united their supporters at the White Cliffs on the

Mississippi and destroyed a boatload of soldiers that Willing sent against them.[133]

Subsequently Governor Chester tried to improve the defense of his province by accepting the offer of John McGillivray, an Indian trader, to raise a militia regiment, with five companies. He made Lyman the captain of one of them as a reward for his "good behaviour in exerting himself in forming the inhabitants into associations," but the scheme was too ambitious for underpopulated West Florida.[134] McGillivray's regiment was disbanded. Only two independent companies remained embodied for the defense of the Natchez district. Naturally Chester chose his most valued officers, Hutchins and Lyman, to command them. The arrival late in 1778 of reinforcements to West Florida, including regular officers, New York and Pennsylvania loyalists, and German mercenaries, made the independent companies an unnecessary expense, and they were demobilized.

The Adventurers had survived the first major challenge which war posed for them, chiefly because Willing's main intention had been raiding rather than occupation. When the Spanish declared war on Britain in 1779 and, from their province on the west bank of the Mississippi, attacked the British possessions across the river, their purpose was permanent acquisition. By that time the Natchez district had a garrison of British regulars, but the Spanish had the advantage of surprise and skillful leadership. When the British commandant at Baton Rouge, Lieutenant Colonel Dickson, surrendered to them, he agreed in the capitulation document to cede not just Baton Rouge but the entire western area of the province, including the Natchez district containing nearly all the holdings of the Military Adventurers. Many of them were outraged that their lands and Natchez, with its stronghold, Fort Panmure, should have been given up without a fight.

When Gálvez, the Spanish commander, depleted his army in the west in order to concentrate on the capture of Pensacola in 1781, they felt the time was ripe for a diversionary repossession of Fort Panmure that would enable Major General John Campbell at the West Floridian capital to regain the initiative from the enemy. Campbell agreed, and the attempt was launched on 22 April. Including Indians, the rebels were 800 strong, but they had no artillery, and Fort Panmure would doubtless have resisted their puny efforts for longer than a fortnight but for a stratagem. The rebels intercepted a message for the fort's commandant, Juan de la Villebeuvre. An addition to it forged by the rebel John Alston indicated that the fort had been undermined. Quantities of gunpowder brought from Pensacola were allegedly in position and due for explosion that very night. The gullible Villebeuvre ran up a white flag over the fort, which was promptly occupied by the triumphant rebel besiegers. Their tenure was short lived. When news arrived that Pensacola had fallen to the Spaniards on 8 May 1781, less than a week after the British victory at Natchez, nobody there believed that Fort Panmure could be retained.[135]

There were two opinions as to the best course of action. Supporters of Captain John Blommart preferred surrender to the Spanish on the best terms obtainable. Others, probably the majority, preferred flight. Of these, about eighty headed north looking for and finding safety among the Chickasaw. They settled on the site of what is now Memphis. Another group of twenty-one headed for the country of the Choctaws, who killed all except two of them.[136] But yet another refugee group, which included the leading Adventurer families, decided on the more arduous alternative of traveling to Georgia, parts of which were still in the king's hands at the time. Savannah was a good 700 miles away as the crow flies, but geography and the presence of hostile Indians and Americans made a direct journey there unthinkable. Instead the refugees planned a circuitous itinerary from Natchez up to North Carolina, then down below the Altamaha, and finally across the colony of Georgia to Savannah.

The entire party was mounted, but progress was slow because it included many women encumbered with infants and, as time went on, a number weak with illness. Their route lay through marshes, canebrakes, and mountains and over bridgeless rivers. They had to pass through the lands of the pro-Spanish Choctaw, who had never forgiven the Adventurers for settling in their traditional hunting grounds. Other Indians were simply rapacious. Two braves followed the party for 500 miles awaiting the chance, which eventually came, to steal two valuable horses.[137] Bears, mountain lions, and wolves too lay in their way. They had little protection from the sickly southern summer, for they had no tents. Lack of time had prevented them from gathering either the provisions or the necessary pack animals for a trek of several months. Their food stock was soon exhausted and they became reliant on wild fruit, berries, and the game they could shoot. Their only compass broke early in the trip, which was a catastrophe.

A heroine of the ordeal was Cynthia Dwight, who traveled with her husband Sereno and their two-year-old daughter Martha. At one point the exhausted travelers came to the Coosa River, which was half a mile wide and, given their condition, insuperable. At last one of the men said that, if anyone would go with him, he would cross the rapid current on horseback to search for an abandoned canoe on the opposite bank. Only Mrs. Dwight responded. His wife's example persuaded Sereno to volunteer as well and, though swept downstream, the three riders swam their mounts across the river. After much searching they found an old pirogue full of cracks. They caulked it with strips of cloth torn from their clothes and brought the remainder of the company over the river three at a time.

If there was too much water in some places, in others there was none at all. At one point, after thirty-six hours without a drink, some of the travelers died of thirst. Toward the end of the next day, when many had totally despaired of survival, Cynthia Dwight again found a reserve of endurance and announced her intention of continuing to walk in search of water until she either found it or could walk no more. Once more her example roused others. Two men and two women offered to join her. Finally, in a low spot between two hills, they felt spongy ground beneath their feet. Digging clumsily with knives, sticks, and hands, a spring was uncovered, and the entire party and their horses were saved from death.[138]

After the refugees reached what they thought was Georgia, they ran into what was described as "the first village of the Muskogee Indians," probably Little Tallassee, the home of the Creek leader, Alexander McGillivray. They were starving, so, in spite of the obvious risks, three of the company were sent forward to ask for food. The Indians were suspicious. Although strongly pro-British, they noted that the newcomers' saddles were such as were used by their enemies the Virginians, for so they called all Americans. Seventy braves surrounded the three scouts, talking excitedly and fingering their knives. Such words as could be understood—"Virginian," "long knife," and "no good"—were unpromising. At that moment rode up a black servant of Colonel McGillivray, who, unfortunately for the travelers, was absent from his home. The black knew both English and Muskogee and, as soon as he understood who the new arrivals were, explained to the skeptical villagers that these were friends. A Creek spokesman said that if these were indeed English they would have a journal; all the English, in their experience, kept journals.

No refugee had in fact maintained any kind of diary. Anything written would do, said the black. Sereno Dwight found an old letter in his pocket and from it pretended to read an account of all the refugees had been doing since leaving Natchez. The black translated this fiction sentence by sentence into Muskogee. The Indians relaxed and relented, giving the travelers the food and the chance for the rest they so desperately needed. The migrants then decided to divide into two groups for the rest of their journey to safety.

One of these groups ran straight into the arms of Americans and was made captive. Members of the other group made their way to the Altamaha River, crossing to its southern bank, where they were less likely to meet revolutionaries, and followed it to the sea. They crossed at the mouth of the river on rafts large enough for both them and their horses and then made their way northward up the coast of Georgia until they reached Savannah. Their journey had taken 149 days, during which they had traveled 1,350 miles. Soon after their arrival they were joined by the group recently made prisoners. For reasons of convenience probably, its American captors had decided to let its members go.[139]

The survivors included Sereno Dwight, his wife Cynthia, their baby Martha, and Sereno's twenty-year-old brother Jonathan. Sereno enlisted in the British army, where he served as surgeon until he was lost at sea off Halifax in 1782. Seven years later Cynthia married a farmer, one of the Northampton Lymans, and died in childbirth the following year. Jonathan became a publican in New York state after the revolution.[140]

Another Dwight, Sarah, though not Sereno's sister of that name, had married a Military Adventurer, Benjamin Day, who had accepted a commission from Governor Chester before the Natchez rebellion. He fled after its failure and safely reached Savannah with his wife, having lost a daughter on the journey through the wilderness. After the revolution Day returned to Springfield, Massachusetts, and resumed his former trades of wool merchant and hatmaker.[141]

The Lymans were even less fortunate than the Dwights. Perhaps as a result of their strenuous journey, both Eleanor and Experience died after arrival in Savannah.[142] The male Lymans who had accomplished the crosscountry trek, Thaddeus, Oliver, and Thompson, all served the British in the later stages of the revolutionary war. Oliver went to Nova Scotia and died insane, Thompson disappeared in the obscurity of the Bahamas, while their elder brother Gamaliel gave up his British army commission at the end of the revolution in order to return to Suffield, where he died penniless. Thaddeus also returned to Suffield, where he lived for eight years before deserting his family to go to New York, where he died in 1812.[143]

Nanachay and other lands granted to the Lymans naturally did not stay in the family. In October 1805 one John Peck of Newton, Massachusetts, tried to recover by legal means the Lyman lands for himself. He alleged that Oliver had possessed 4,000 acres of the main family plantation and had inherited a half share of the land apportioned to his now dead brother Thompson and sisters Eleanor and Experience, which totaled another 2,500 acres. Peck also alleged that on 11 October 1805 he paid Oliver $13,000 for his claim to any land in Florida, which he authorized Robert Williams, the new governor of the Mississippi Territory, to take up for him.[144] Whatever the result of Peck's efforts, it is satisfying to learn that, after all the expenditure of gold, life, and spirit, at least one member of the Lyman family made a little money from the land.

The question of the loyalism of the Military Adventurers is interesting. Superficially it might be assumed that, emigrating from Connecticut when they did, on the eve of the revolution, they were attempting to escape the turmoil of the rebellion which was so obviously imminent. This view would suggest that, without necessarily being enthusiasts for royal authority, they respected it; from the beginning they sought royal approval for their plans and were certainly not keen revolutionaries. Some credence is given to this interpretation by their towns of origin. Connecticut was divided geographically in an unusually clear manner in the great debate on American rights preceding the revolution. Adherents of Governor Thomas Fitch, supporters of parliamentary authority over the colonies, came from west of the Connecticut River, which bisects the colony, where, and not by coincidence, Anglicanism was strong.[145] On the other hand, partisans of Governor Jonathan Trumbull, Sons of Liberty politically and New Lights in religion, came from eastern Connecticut. Of forty-one emigrants identified by Matthew Phelps by place of origin, thirty-six came from west of the Connecticut River. The other five all came from the same town, Lebanon.[146]

On this evidence alone it might be assumed that the vast bulk of the emigrating Military Adventurers would be cool to revolutionary ideals and favorable to Governor Fitch. For several reasons, however, any such assumption would be oversimple. For one thing, Fitch was hostile to schemes of westward emigration. In 1762 he antagonized the Susquehannah Company by proclaiming a ban on western settlement by its members.[147] Since many of them were also shareholders in the Military Adventurers, many of the Mississippi-bound emigrants would surely have despised Fitch. An egregious example was Hugh Ledlie, treasurer of the Adventurers,[148] a migrant to the Mississippi in 1774,[149] a strong critic of Fitch for his support of the Stamp Act,[150] and a leader of the Sons of Liberty in Connecticut.[151] Other leading Adventurers were more actively revolutionary even than Ledlie. Of the six committee men that the company sent to explore the Mississippi in 1773, at least three became generals on the revolutionary side during the war. Major General Israel Putnam was the most famous, but his cousin once removed, Rufus, and Roger Enos both became brigadier generals during the conflict.

On the subject of loyalism the evidence of Matthew Phelps is suspect. He wrote in 1802 for an American readership and consistently downplayed the existence of loyalists in West Florida. His description of the settlers in the Natchez district as "well-disposed to the American cause, almost universally" should be treated with caution. Nevertheless the large number of hunters— eighty—who joined James Willing, of whom Phelps had a low opinion as doing "incalculable injury to the American cause," does not suggest deep devotion to king.

Of course there were such devotees among the Adventurers, most notably the Lyman family. The old general's several sons had a chance to fight for George III and did so. Loyal too were the Dwights, and some of the worst sufferers in that family were those who had to stay in New England.[152] They were jailed or had their property destroyed.[153] The price of loyalty was high too for the family of Timothy Hierlihy, who had served under General Lyman in the Seven Years' War. Although he had accompanied one of the earlier Adventurer expeditions to the Mississippi, he had not settled there but had returned to Middletown, Connecticut, which he left in July 1776 to join tories on Long Island, where he was commissioned in a loyalist unit. His wife, Elizabeth, and their nine children were left in Middletown and suffered severe economic distress, but the Connecticut assembly rejected Mrs. Hierlihy's petition to be allowed to join her husband in New York.[154] By the end of the war, Hierlihy had risen to the rank of lieutenant colonel in the Nova Scotia Volunteers, in which unit his sons Timothy and John were a captain and an ensign.[155]

An interesting piece of evidence relating to loyalism was a memorial addressed to Alexander Dickson after he had surrendered Baton Rouge to the Spanish in 1779. It may be remembered that, in addition to Baton Rouge, he gave up the whole of the western portion of the colony, including the Natchez district. The memorial was signed by fifty-nine of its inhabitants. Strong loyalists writing such a memorial might have been expected to deplore Dickson's swift surrender of Natchez without giving British subjects there a chance to fight. On the contrary, these signers thanked Dickson in fulsome terms for his conduct, implying that they much preferred peaceful transfer to Spanish rule to the necessity of fighting. They included five of the Military Adventurers. The names of Elijah Leonard, Benjamin Day, Timothy Hotchkiss, and Jacob Harman are not unduly surprising, but that of Thaddeus Lyman certainly is, because of his bellicose role in the Natchez district and because he was subsequently recognized by the British authorities as having a right to a place on the official list of loyalists.[156]

Numerous Military Adventurers came from elsewhere in New England, but the bulk of them came from Connecticut. Fifty of the sixty-four emigrants, exclusive of wives and children whom I have identified, came from that colony. Not counting the highfliers of the exploring committee already mentioned, the names of no less than thirty-three of the fifty reappear in the rolls and lists of Connecticut men who served the American cause in the revolution, three as captains, one as a lieutenant, one as an ensign, one as a sergeant major, twenty as private soldiers, and the rest in other capacities, such as supplying pork to the army. In view of the number of emigrants who died, this figure is extremely high. It is hard not to believe that, in spite of the professions of loyalty to the crown and of abhorrence of the revolution many of them had made when applying for land, most of the Adventurers were not loyalists.

There was no very good reason why they should be loyalists. If economic factors were important—and surely they were—then the British government had subjected them to much frustration. Initially it had issued a proclamation which apparently entitled them to sizable grants if they had served in the Seven Years' War. Many of the Adventurers with such service discovered that the proclamation was interpreted as not applying to them. Then they paid for their leader to go to London to obtain a different type of block grant for them all. The British repeatedly raised his expectations but in the end denied land for his company. The Adventurers went ahead with their emigration plans in any case, and once the process was well under way, the British crown put a restraining order on almost all land grants. The ban was relaxed in 1775, but even then the legalities and the fees required were formidable. When they finally settled, the British provided them with inadequate military protection, with the result that they fell to James Willing in 1778 and the Spanish in 1779. Most of them then went home to New England, which I believe does more to prove that they were not loyalists than does subsequent service in the American armed forces. Few indeed put down roots along the Mississippi. The Spanish census of 1792 reveals that only six of the Adventurers were still living in the Natchez district. In remaining, these six do not necessarily demonstrate any particular national commitment. One of them, Bernard Lintot, turned his coat almost as readily as most men change their socks.[157]

So if loyalism did not drive the Adventurers to the Mississippi and the British did so much to disappoint their economic expectations, we must ask if their motivation was economic in the first place. The short answer is yes, because the economic prospects in Connecticut were even more dismal than in the wilderness. The roots of economic discontent in Connecticut went back a long way, and it should be noted that both the Military Adventurers and the Susquehannah Company were founded several years before there was a quarrel between the English king and his American subjects.

The basis of the economic hardship was the large migration into Connecticut from Massachusetts in the early 1700s, which had created acute land hunger and, by the standards of the day, overpopulation. Oscar Zeichner has analyzed the underlying economic reasons for support of the Susquehannah Company. Most of them would apply equally well to the Company of Military Adventurers. Connecticut was, next to Rhode Island, the most densely populated North American colony: its population increased by 50,000 between 1762 and 1774. Its soil had been worn out by incessant cropping and backward farming methods. It was difficult for an independent farmer to make a living from perhaps twenty acres of impoverished soil and not easy to obtain more. Much of the usable farmland was either of unreachably high price or was held by absentee landlords for speculative purposes or for rent. It is understandable that many small farmers, faced with the alternatives of making a decreasingly good living from their own land in Connecticut or farming someone else's acres there on the one hand or, on the other, of acquiring comparatively large acreage for themselves as a gift of the crown through emigration, preferred to move. The emigration from Connecticut in the period 1762 to 1774 was estimated by Benjamin Trumbull at something over 2,000 a year.[158] If the Connecticut emigrants to West Florida survived, they could expect to own on the banks of the Mississippi some of the richest soil in America and never again to worry about providing for their children.

Also in the short term the emigrants had not needed to worry about choosing sides in the quarrel between Britain and her colonists in the years following the Seven Years' War. They could afford the luxury, welcome to many no doubt, of standing aside. Nevertheless there came a time when aloofness was impossible, although the moment for choice was delayed longer in West Florida than elsewhere. When this happened, clearly many New Englanders preferred to face the martial music back home.

How can we summarize the efforts of the Adventurers, apart from their illustrating, as Cecil Johnson noted, the impulse toward westward expansion? First, theirs was the largest single organized scheme for settling in West Florida. Exact numbers are hard to compute, but Silas Deane was guilty of little exaggeration when he wrote in 1775 that several hundred young men from New England had gone there with their families, and more followed. After all their suffering and the revolution were over, however, and when Matthew Phelps on behalf of himself and other survivors of the company sought compensation for the lands they had lost in 1804, when the soil on which the Adventurers had camped had become part of the United States, Congress flatly refused.[159]

Of course their story is a tragic one, leaving an overwhelming sense of loss and waste. Until their arrival West Florida had been, on the whole, a magnet for second-raters, but people like the Dwights and Lymans were of the very best. Whether education, ability, or moral fiber is the criterion, they were superior individuals. And yet their fate was sickness, impoverishment, persecution, and death; sadly so, since with luckier timing their ambitions might easily have been crowned with abundant success. If, say, Phineas Lyman had wasted only three years instead of nine in fruitless political lobbying, a strong community might have been established in West Florida which could have altered the course of history—in which direction is hard to tell. Possibly the Spanish conquest would have been averted, in which case the province might have remained British or, alternatively, become a fourteenth state named perhaps Georgiana or Mississippi.

The Lymans and Dwights, however, were not typical. Perhaps of more significance were the numerous undistinguished families accompanying them who left hard times in New England to find even harder ones on the Mississippi and so went back home. Independence did not bring them prosperity, if examination of Connecticut probate records is any guide. Most of them died leaving little more than a few clothes, a few sticks of furniture, and, sometimes, a few acres and a modest dwelling. Their sufferings, courage, and versatility were probably as great as those of the Dwights and Lymans, but they had less leisure and fewer literate family members to document their odyssey.

8

Conclusion

From its beginning as a British province, West Florida was beset by daunting economic problems. Some were surmountable and some not—not, at least, in the short term. Despite a considerable amount of well-publicized optimism about the colony's economic prospects—not all of it ill founded, for West Florida's potential was undoubted—its promise was unfulfilled.

The economic difficulties may be summarized as sixfold: a persistent trade imbalance, high costs, and four shortages—of markets, of capital, of labor, and, above all, of time. In the discussion that follows it should not be forgotten that for Britain the economic significance of West Florida was not its only or even its chief importance. Its harbors on the Gulf of Mexico which could mean Royal Navy bases, its proximity to the possessions of the traditional Bourbon enemies, and its extension into the interior, which facilitated influence over the western Indians, all gave it high strategic value.[1]

Nevertheless that value could not be realized unless the province attracted settlers and in their mind the promise of economic betterment was uppermost. Except for poor wretches like Montfort Browne's imports, who perhaps were not making a living in Ireland, the promise of merely making a living would be insufficient. The prospect had to be of a better living; hence the offers of abundant free acres, the enticing talk of trading at huge profit with the dons, and the discussion of West Florida's potential (never to be realized) for producing exotic items like silk, wine, brandy, cochineal, coffee, and quassia, which were obtainable nowhere else in quantity in the British empire. Free land, however, is useless unless it is worked, and worked land unprofitable unless there is a market for its products. Trading with foreign countries is difficult without benevolent governments, and rare products difficult to perfect in an untried soil and climate, especially by inexperienced optimists. The disappointments of early settlers in West Florida were made known in the press on both sides of the Atlantic, and the immigrant batches tended to be in dozens or at best hundreds rather than the thousands needed before the colony could hope to be self-sufficient; and until self-sufficiency was achieved, a trade imbalance was inevitable.

Lieutenant Governor Montfort Browne shrewdly analyzed the chief economic problems of his province and how they were intertwined some five years after its foundation:

> The price of labor is so excessively high and the number of planters so few that this delightful province, so capable of contributing to the aggrandizement of the mother country, has not since its first settlement been able to produce any commodity that can answer as a staple to exchange with neighbouring colonies for the mere necessities of life and the cash that would otherwise circulate among ourselves we are obliged to remit in exchange.[2]

In fact there was a staple, the same commodity, furs and skins, that the French had exported from their part of the province before it changed hands in 1763. Browne remembered it some weeks later when he again wrote to the plantations secretary, but the point he made was essentially the same. Annually, he alleged, Mobile, where the fur trade was centered, was exporting 150 hogsheads of peltry, whereas West Florida was simultaneously importing between £80,000 and £100,000 worth of manufactures and other articles from Great Britain.[3] In short, there was an enormous disparity between what was being bought and what was being sold back to help pay for it.

It is difficult to know precisely how many skins would fit into a hogshead, a container of variable size which, at different periods, contained as little as 63 gallons or as much as 140; but it does seem possible that 150 hogsheads might roughly equate to the 4,636 skins which were officially reported as having been exported from Mobile in August 1766, the month when Browne made his estimate.[4] If so, the skins weighed approximately 11,590 pounds and, at 40 sols per pound, the average price in the period, would be worth 463,600 sols (about £5,675, or $24,403). The figure would thus confirm Browne's point about the low value of West Florida's exports in 1766.

The amount of peltry exported rose greatly during the British period. Peltry in fact remained the foremost West Floridian export, but even in the bumper year 1774, the peltry exports to Britain amounted to only about £20,000.[5] Even if we double that figure to take into account the unrecorded peltry sold in New Orleans, the figure comes nowhere close to the amount spent on imports, which, from Britain, were estimated in the 1770s at an annual £97,000.[6] Although peltry continued as the main export, by the mid-1770s (after which war diminished the trade) West Florida was also exporting logwood, dyewoods, tobacco, indigo, and staves. These were on the rise but still distinctly minor items. Tobacco imports to Great Britain, to take but one example, totaled a mere 6,062 pounds in 1770 and none during the three succeeding years.[7] Cluny and Entick estimated West Floridian imports during the same period as worth £63,000 a year, a figure that was probably an exaggeration but one that suggests the continuance of a significant, if closing, trade gap.

Some conclusions may be drawn from the export and import figures collected by David MacPherson, who first published his *Annals of Commerce* in 1801. MacPherson probably had access to a more comprehensive range of trade statistics than Entick and Cluny, for, unlike them, he published a full range of trade figures for the Floridas for every year of their existence.[8] Even though he lumps both colonies together, the sums he cites are much lower than those for West Florida alone of other writers. Where Montfort Browne estimated imports from Britain to West Florida at £80,000, MacPherson's figure for 1768 is £29,509.[9] A collection of West Florida petitioners thought the exports of their province to Britain in 1778 were worth £200,000.[10] MacPherson's figure for that year is £23,804. And whereas Entick thought that during the early 1770s West Florida imported an average of £97,000 and exported to Britain £63,000 in value a year,[11] MacPherson's figures for that period are as follows:

	Floridian exports to Britain (pounds sterling)	Floridian imports from Britain (pounds sterling)
1770	21,856.11.11	66,647.9.11
1771	15,722.17.6	40,458.2.9
1772	7,129.13.6	51,502.7.2
1773	22,335.19.5	52,149.14.4
1774	21,504.19.6	85,254.7.10

His figures are so different from those of other writers that they suggest great caution in accepting any contemporary estimates for West Florida's trade. All surviving estimates, however, do agree on the existence of a sizable trade gap in Britain's favor, even though MacPherson's do not point, as Cluny's and Entick's do, to a gradual closing of the gap. Actually, Mac-Pherson's figures for the subsequent war years do indicate that, in the final years of the conflict, the Floridas were sending more to Britain than Britain was selling to the Floridas, but this is an aberration easy to explain.

From 1779 British merchants would be reluctant to dispatch goods for which, if known Spanish successes in Florida continued, they would never receive payment. Lower import figures, then, did not mean diminished demand. The normal pattern in the British period, including the war years, was that West Floridians wanted far more from Britain than they could possibly pay for with exports.

West Florida was, therefore, a debtor province, in which it was no different from older, longer-established, and more prosperous colonies. At the time, the trade balance of the North American colonies taken as a whole was estimated as being annually £1 million in favor of Britain.[12] This estimate agrees fairly well with that of twentieth-century economic historians Gary Walton and James Shepherd, who put the gross deficit for the years 1768–1772 at £1,120,000.[13] Older colonies bridged the gap in the balance of payments with the sale of shipping, shipping services, insurance, and other "invisible" items and by piling up debts which, provided the interest on them was regularly paid, did no great harm economically, although it was made an important political grievance.

West Floridians became debtors too, but the province had other means of alleviating the gap in the payments balance. Selling to Britain was not the only source of income. The British government, not the colony, for example, provided for its civilian and military officials, whose pay and allowances were spent in the province. The salaries of the civilian officers, together with grants for Indian presents, generally hovered around £5,000 annually,[14] although in 1773 they totaled the extraordinary figure of £7,900.[15] In addition £315 a year was allowed for the support of a government schooner. Of more consequence was soldiers' pay. In other colonies redcoat garrisons were a grievance. In West Florida, where they offered security against the attacks of various potential enemies and greatly improved the payments imbalance, they were not. In 1771 the provincial agent in London estimated that the pay alone for the two garrison units would come to £33,078.2.6 in the coming year.[16] Admittedly the percentage of this amount which went to officers would not all be spent in the colony, but another military expense which benefited the province was that part laid out by London for soldiers' rations spent on locally produced food. For example, two months' rations in 1767-1768 cost £1,587.6.6. The most important item was *fresh* beef, which the agent Joseph Garrow must have bought in West Florida.[17] Added to this expense would be the cost of building and repairing military structures in the province and of extraordinary military duties. The survey trip that engineer officer John Cambel took to the rivers and lakes of West Florida in 1767 at a cost of £421 is an example of this type of expenditure.[18] Sailors stationed in Pensacola or on shore leave would also benefit the colony's economy. Management of the Indians meant that London supplied money that was spent in West Florida. William Burnaby alone spent £200 on entertainment at the Pensacola Congress in 1765,[19] but the entire expenditure of the Indian Southern Department for the nine months during which that congress took place, the vast bulk of which was spent in West Florida, was £18,770.19.0 in New York currency or £10,172 sterling.[20]

Another source of income for the colony was the illicit and therefore unrecorded amounts brought in from sales to foreign colonies, of which Louisiana was the most important. In 1776 Francisco Bouligny estimated that Louisiana's annual trade in indigo, peltry, and lumber was worth a total of 480,000 Spanish dollars (or £111,628 sterling) of which $465,000 (£108,140) fell into British hands.[21] This figure seems inordinately high, and almost certainly Bouligny was blaming the British for trade which in fact benefited other foreigners. Jacob Blackwell believed that most of the twenty vessels plying the Mississippi in 1775 illicitly flew English colors. Manned by French crews and bearing French cargoes and owned by French merchants of Hispaniola, sometimes, for an appearance of legality, they employed an English commander, and occasionally the vessels carried spurious English papers, but mostly they sailed without any. A genuine English snow regularly loaded up with skins and indigo below New Orleans, but they ended up in France, not England.[22]

So, although there is no question that Britons did benefit from selling in Louisiana, they benefited far less than Bouligny seems to believe and, although they alleviated the trade imbalance, their activities probably fell far short of fully redressing it. Nevertheless the Spanish milled dollar and the ryal and, to a lesser degree, the livre and the sol were coins in commoner use in West Florida than guineas, shillings, and pennies. Even the British garrisons there were paid in Spanish money.[23]

The shortage of ready money, which from the beginning was a nuisance in West Florida, combined with the British Parliament's ban on issues of paper currency, led to extensive use of credit, which took several forms, a particularly frequent one being the bill of exchange, which in turn led to high prices. The higher prices were understandable, because a bill of exchange drawn on a reputable London firm or the British treasury could be honored only in Britain. The process took time, months at best; and there was the real possibility that a bill would be refused and actual payment would then be further delayed or even denied. Who then, prepared to accept a bill of exchange as payment, would undercharge? Despite the prevalence of bills of exchange, ready cash was always preferred. In asking the governor of Virginia for money in 1770, John Stuart begged for gold rather than paper, since it would not be accepted unless he allowed a 3–5 percent discount on its nominal value.[24]

Another circumstance pushing up prices in Florida was shortage. Initially nothing, not even food, was locally produced. Newcomers were shocked by food prices, and complaints about them found their way into the British and colonial press, no doubt helping to deter doubtful would-be emigrants to Florida. I list some prices from the *Scots Magazine* of January 1764 and 1765, a letter of June 1765, and the *Georgia Gazette* of 21 June 1764; in parentheses I give the prices of comparable items in the period November 1772 to February 1773 (note that there were twelve pennies in a shilling and twenty shillings in a pound): a chicken, five shillings (sixpence); a turkey, fifteen shillings (one shilling); beef, fivepence a pound (a penny halfpenny), and pork, tenpence a pound (twopence).[25] By the 1770s food prices were dramatically lower, because self-sufficiency had been achieved in many but not all foodstuffs. The province was not especially suitable for rice, although a little was grown.[26] Whites used rice only for puddings and soup, but cracked rice was a staple of the slave diet and of the British soldier's daily rations, and South Carolina exported hundreds of barrels yearly to West Florida in its early years.[27] By 1772 West Florida itself exported 192 barrels of rice to Britain and a similar amount to another colony, so, even in this commodity, the colony was probably moving toward self-sufficiency.[28]

Nonedibles were also regularly imported and inevitably were comparatively high in price. Nearly all of them had to come huge distances by sea, with the freightage and insurance added to the basic cost as well as the price boost produced by scarcity. A correspondent was particularly enraged in 1765 at having to pay thirty dollars (or seven pounds) for a hat which would have cost three shillings in England. Even in the 1770s, in the remote Natchez region at least, there was discontent at having to pay twelve dollars for a pair of thin blankets and six bits for a yard of material.[29] To judge from the prevalence of homespun and woven clothing mentioned in various accounts, it is probable that the price of imported clothes remained high throughout the British period in time of peace, and the extraordinary shortage of imports in the later stages of the war—MacPherson's figure for 1780 is £16,446 and for 1781 a mere £4,707—would certainly have perpetuated high prices.[30]

Inflation may be suspected from the escalation of fees charged by officials concerned with the acquisition of land. The governor, provincial secretary, attorney general, and surveyor general all charged for their services. In the 1770s Governor Chester and his secretary and friend, Philip Livingston, hiked their fees significantly. In 1765 the governor's fee on the sale of 1,000 acres of land was set at $8. In 1772 Chester raised it to $28, and in 1778 to $80. Livingston was more extreme. For similar acreage the secretary's fee in 1765 was $4.2 ryals. In 1772 it became $51.4 ryals, and in 1778, $106.2 ryals.[31] Almost everybody who settled in West Florida acquired land, and thus the effect of these increases would have been widely felt, but they probably reflect rapacity rather than inflation: throughout the period the attorney general charged the original nominal fee. The fee of the surveyor general, at $153 for 1,000 acres, was the biggest for the would-be planter, but it too did not change. Then again, as we have seen, the price of slaves remained constant, and the price of land as well in most areas did not escalate, because there was always an abundance of free land available. There was stability, moreover, in the price of Indian goods, while the legal interest rate, at 8 percent, seems to have remained constant and is one that does not suggest inflation before the outbreak of war.[32]

One of the most trumpeted advantages of Britain's acquisition of West Florida was the expected access it would give British goods to foreign markets. From British customs summaries Charles Mowat found that 115 vessels left Pensacola in the four years 1769 through 1772.[33] The comparable figure from the entries listed below in Appendix 1, which are culled mostly from newspapers, comes close to Mowat's. The data in Appendix 1 show that most shipping came and went to other British ports. The one Spanish port which is at all noticeable is New Orleans, and Louisiana was easily the most profitable Spanish market for West Floridians. For various reasons, of which a lack both of support from the British government and of Spanish complaisance are salient, the anticipated development of markets in the provinces of Mexico never took place, and trade with them remained minuscule, while commerce with Cuba and Spanish Hispaniola was nonexistent.

Another factor inhibiting trade with the Spanish empire was a shortage of capital, and there are reports of Spanish visitors arriving with longed-for bullion, only to find nothing to spend it on in the empty storehouses in Pensacola.[34] West Floridians put a large percentage of the capital available in the province into slaves rather than merchandise. Hence Pensacola and Mobile never turned into booming entrepôts, and a number of the more aggressive British and American traders preferred to transfer their premises to Spanish New Orleans. The coming of war did open up an unexpected market for West Florida products, particularly lumber, in the British West Indies, which were cut off by war from New England, their traditional source of supply. Had West Florida remained a British colony after the revolution, as the West Indies did, no doubt that market would have continued to exist and perhaps would have developed but would not have matched in opulence the trade with the Spanish empire anticipated by the dreamers of the 1760s.

West Florida was vast, but its population expanded with disappointing slowness, and there was a chronic labor shortage. White indentured servitude was noticeable but uncharacteristic. In a fluid society full of a variety of opportunities where natural abundance made surviving in idleness an option, the apprentice cobbler would not stick to his last. "White servants will never turn to account," wrote Romans.[35] Actually there was little demand for white indentured servants in West Florida, as Montfort Browne's poor deluded Irishmen found to their grief. The reasons were twofold. One was a general belief that whites were constitutionally unable to work well in the hot southern climate.[36] The more important reason was economic: whites were not profitable compared to the alternatives. The prime alternative was not the Indians of West Florida, although the Choctaw was known to be industrious and to be ready, as other local Indian tribes were not, to do menial work for whites.[37] Indeed there were some few enslaved Indians in West Florida, but the more obvious alternative was the black slave. He or she cost less to feed than a white servant, could be worse housed, had fewer legal rights, and could be worked for longer hours and for more years than a white. The greater desirability of slaves was obvious to West Floridians like Daniel Hickey, who wrote to his employer about Browne's plantation, "There is nothing wanting to make it a pretty place but a few negroes for it is so expensive to keep many white people upon it; they won't put up with what negroes would live upon."[38] Provided they survived, black slaves could return a quick profit. Admittedly a mature male field hand would require a heavier investment than in the northern colonies, about $350, yet on the Mississippi he was estimated as likely to make $100 a year for his master.[39]

Of course the British did not introduce slavery to West Florida. When they arrived in 1763 they found several established French families like the Rochons of Mobile who had dozens of slaves on their plantations.[40] But the British did enormously extend slavery; it was an obvious expedient to combat the acute and chronic shortage of manpower. Proof of that shortage is that during the war the West Florida government resorted to the desperate and, what in other colonies would have been regarded as dangerous, expedient of arming the slaves.

Perhaps of more importance in denying British West Florida the prosperity so often prophesied was shortage of time. It normally took more than eighteen years—the time which elapsed from the arrival of the first Britons to the Spanish conquest of the province—for a developing colonial economy to become viable, even if the last six years, as was the case with West Florida, were not warped by war. The sole colony established by Britain in North America in the eighteenth century prior to the acquisitions of 1763 was

Georgia. Despite obvious differences such as the initial ban on slavery there, Georgia was enough like West Florida in size, latitude, climate, and soil for comparison to be plausible. Romans noticed that, for over twenty years after its foundation in 1732, Georgia's economy languished but that in the following two decades her exports boomed. To back his assertion he cited statistics compiled by the comptroller of customs at Savannah, according to whom Georgian exports rose in value from £15,744 in 1755 to £121,677 in 1772.[41] These figures are larger than the estimates of twentieth-century economists. For the same period the United States Department of Commerce gives the growth as from £4,437 to £66,083.[42] Walton and Shepherd come slightly closer to Romans's estimate, and they assert that in the years 1768–1772 Georgia's exports per head of population were higher than those of any American colony except South Carolina.[43]

If we look more precisely at deerskin and indigo exports—which were also West Florida's most promising products—it may be seen that Georgia peltry exports rose from 49,995 pounds in 1755 to 284,840 pounds in 1770 and that indigo exports bounded from 4,508 pounds in 1755 to 22,336 pounds in 1772.[44] There is little reason to doubt that a comparable expansion could have taken place in similar West Floridian exports. After twenty-two years of existence as a colony, Georgia for the first time enjoyed successive years in which its exports exceeded its imports.[45] If West Florida had been given the luxury of a comparable period of time, it is entirely possible that it too could have achieved a favorable trade balance in the 1780s, although Georgia had one important asset, a good port on the Atlantic, that West Florida would always lack. This deficiency would not have been completely offset by Pensacola's easier access to Jamaica, but in other respects, given time, the colony of West Florida might have equaled Georgia's development.

In the early 1770s immigrants of varying caliber, of whom some were excellent, came pouring into West Florida by river and sea from other colonies in sizable numbers. By then settlers knew where the richest land was located, and through trial and error they found that promising products like indigo, tobacco, and timber derivatives could flourish in West Florida, and could supplement peltry, the existing staple. It would only have been a matter of time before the colony became self-sufficient in rice and produced and sold cotton in quantity to feed the still-growing appetite of Britain's textile industry.

For a colony as an entity to thrive or to fail or, as West Florida did, to develop despite setbacks toward an unfulfilled prosperity, does not necessarily reflect the fortunes of individual colonists. Always in times of misfortune some will flourish, and in boom times invariably some will fail. There is not

much doubt that in West Florida disappointment was commoner than afflu-
ence. After the chimera of trading with Spanish colonies faded, a great many,
perhaps most, West Floridians pinned their hopes of prosperity onto the
prospect of appreciation in land values. More to be pitied than Philip Livingston, who, clearly as a speculation, sedulously amassed tens of thousands
of acres which yielded him nothing, were the numerous small planters who
found that all the years of arduous clearing and building and improvement
had been spent in vain. After the Spanish conquest their property was
worthless, unless they were prepared to live under the flag of Spain. It is hard
to understand the almost universal abhorrence for that alternative, but few
planters stayed after the British surrender. It is hard not to believe that more
would have remained if they had been making comfortable livings from the
land. Even if a colonist did not live off the land, making a fortune in West
Florida seems to have been impossible. The province produced no nabobs,
men who—curious phenomenon of eighteenth-century Britain—had been
able to pursue careers in India so lucrative that, upon retirement in their
native land, they could lead lives of ostentatious splendor. Nevertheless certain
classes of inhabitants did a great deal better than merely survive in West
Florida.

A person was fortunate who possessed skill in an occupation badly needed
in the province. William Aird was so blessed. He was a house carpenter. In
other colonies or Britain he would have ranked merely as a respectable and
useful member of society. In West Florida, where the need for wooden houses
in quantity was urgent, he was a grandee. His own "commodious" house
was big enough to accommodate all the government offices, the legislature,
and the law courts too.[46] In 1767 he paid more tax on more slaves, nineteen,
than anyone else in the province.[47] With his brothers he owned 3,000 acres
of rich plantation land on the Mississippi,[48] and he repeatedly represented
Pensacola as one of its two members in the colony's legislature.[49]

Planting too could be profitable, and one suspects that the successful
planters were among those least reluctant to live under Spanish rule. William
Dunbar is an obvious example, but Charles Parent, who had a six-room
house on one of his several plantations, numerous cattle and slaves, and a
sizable tarmaking sideline, is another.[50] After one of his houses was burned
by the British, he enlisted in the Spanish militia.[51] Major Robert Farmar died
in 1780 and thus did not have the chance (which he probably would have
refused) to serve Spain, but he amassed an estate worth $35,000 under the
British flag in West Florida.

Although numbers of traders transferred their businesses to New Orleans,
those who stayed in West Florida could also prosper. Samuel Steer turned

temporarily to trading between Jamaica and the Mississippi after his plans to be a planter were ruined by Willing's raid in 1778. He stayed on after the Spanish conquest and continued to do well. When his considerable property was inventoried in 1793, he had forty slaves.[52] Perhaps a less diluted example of success through trading is provided by the firm of John McGillivray and William Struthers of Mobile. Their account books have not survived but, apart from their engagement in the Indian trade and their numerous appearances in the provincial legal records as creditors to more needy fellow subjects, we know that, when obtaining 3,000 acres on the Mobile River in May 1770, they alleged in support of their application that their company owned forty slaves, in themselves a capital asset worth, at a conservative estimate, over $10,000.[53] They were clearly men of substance, perhaps the most eminent traders of Mobile, for both represented Mobile in the provincial legislature, simultaneously in the case of the 1772 session.[54] The business of John Fitzpatrick, the trader of Manchac, was strong enough to survive Willing's depredations. Like his fellow Manchacker, Daniel Hickey, Fitzpatrick was an Irish Catholic, and perhaps in part because they were Catholic, they both stayed on after the arrival of the Spaniards. When he died Fitzpatrick's property was assessed at $7,465.[55]

Trading in slaves was particularly profitable, if writers like Romans are to be believed, although it could not be practiced on any scale without considerable preliminary investment. Slaves were profitably bought and sold as a secondary activity by the likes of Dunbar and Fitzpatrick, but it seems to have been the main business of the Comyn family, which boasted that, at £15,000 ($64,500), theirs was the best-capitalized firm in West Florida.[56] They worked in conjunction with McGillivray and Struthers and used their brig *Africa* to supply Fitzpatrick, among others no doubt, with blacks. They were not ostracized for engaging in a distasteful business. On the contrary, Valens Stephen Comyn was a member of the West Florida council, and Governor Chester wrote with respect of his brother Phillips, another councillor, as "a gentleman of good sense and property here."[57] Slavery was ubiquitous in West Florida but so was the thirst for rum, and judging from the number of licences to retail it that were issued, many West Floridians of no great eminence made a living from selling it. In 1768, long before the population boom of the 1770s, fourteen individuals paid for such licences. It seems that a visitor to West Florida would have had a good choice of taverns along the waterfront of Pensacola or Mobile.[58]

In general the surest route to prosperity seems to have been employment by the British government, either directly or by obtaining a British government contract, of which one of the more lucrative was perhaps the supply of rum to the army garrisons, a privilege enjoyed at one time by the firm of John Stephenson and Arthur Strother; in under three months in 1779, they grossed £2,456.14.6 ($10,564).[59]

The best preferment which British employment could offer was the governorship, for which the salary was £1,000 ($4,300) a year. Lesser crown appointments were also much sought after, not just for the salary but for the fees payable to government officers in most legal and many commercial transactions. For this reason the job of the provincial secretary, who had to approve or record most of them, was particularly desirable, and Philip Livingston, Chester's secretary, probably supplemented his £150 salary several times over with fees. So, undoubtedly, did the surveyor general, whose nominal income was only £120. On the other hand, applicants were scarce or unobtainable for the posts of parson and schoolmaster, who received annually £100 and £25. The provost marshal, although receiving the same salary as a parson, financially speaking had an enviable job, since he could collect substantial fees under the terms of the slave code.

Thanks to the comparatively small legitimate maritime traffic of West Florida, customs officials probably could not live comfortably on their small salaries, even though they were supplemented by fees, unless they engaged in additional money-making activities. The egregious example is Jacob Blackwell, the collector of customs at the port of Mobile. Among his other roles were member of the West Florida council, stamp distributor for West Florida, slave dealer,[60] the major reexporter of rum from West Florida in the late 1760s,[61] and realtor. The acreage and the amounts of money in which he dealt were, by West Floridian standards, large. For instance, on 1 August 1770, jointly with John Scott, he borrowed £3,001.15.0 ($12,908) from Prideaux Selby[62] and subsequently, when short of money in 1773, sold six tracts of land totaling 10,000 acres.[63] His most certain source of income was as agent of the London merchant Edward Codrington, who had secured the contract to supply rations to the British troops in West Florida.

Nevertheless, however possible individuals may have found it to make a good living in West Florida during the British regime and however bright the economic future of the province may have been if that regime had survived the revolution, politically its future would have been more questionable. Canada, it is true, continued after the revolution as a loyal colony without a major rebellion against British rule until the 1830s, but a high percentage of its population was French, heirs to a different tradition from the immigrants to West Florida. Probably only a small minority of the immigrants who came into the more southerly colony in the 1700s were inveterate tories, although the breed undoubtedly existed. Most of them came for economic reasons, and although numbers asserted their toryism after 1775, no doubt often, as in Bernard Lintot's case, it was in order to claim bounty land. In this connection, the large number of Connecticut emigrants of the Company of Military Adventurers apparently prepared to live under British rule who, when their venture collapsed, returned to serve in revolutionary armies, is significant.

As the new American republic in succeeding years demonstrated viability, as new waves of backcountry settlers continued to arrive (not from other colonies any more but from states), as the British government, having changed perhaps its personnel but not the principles upon which its colonial policy was based, continued to annoy with delays, parsimony, and vetoes, and as the Indian and Spanish threats weakened, it is quite possible that West Florida would have become one or two of the United States by the end of the eighteenth century.

Appendixes

Appendix 1
Maritime Traffic, 1763–1778

Year	Voyages to and from West Florida			Voyages to and from New Orleans and Mississippi[a]
	Pensacola	Mobile	Total	
1763	16	8	24	—
1764	62	31	93	1
1765	94	14	108	1
1766	55	7	62	4
1767	60	3	63	13
1768	43	—	43	9
1769	38	1	39	3
1770	31	1	32	6
1771	35	—	35	4
1772	17	—	17	7
1773	23	—	23	19
1774	33	1	34	22
1775	38	—	38	15
1776	66	2	68	—
1777	30	—	30	—
1778	36	—	36	3

Note: Dashes indicate that no voyages were recorded.
[a]Voyages made to and from these settlements from American ports.

Sources. Sources for a comprehensive list of voyages do not exist, and the volume of maritime traffic was no doubt greater than the figures given above would suggest. The main sources used here include colonial newspapers; Account Book E of Philip Livingston in the West Florida Papers of the Library of Congress; a naval officer's shipping list in the Georgia Historical Society, Savannah; *Naval Documents of the American Revolution,* vols. 1–4 ed. William Bell Clark, vols. 5–8 ed. William James Morgan (Washington, D.C.: U.S. Government Printing Office, 1964–); and K. G. Davies, ed., *Documents of the American Revolution, 1770–1783,* 21 vols. (Shannon: Irish University Press, 1972–1981). Warships, droghers, and packet boats have been excluded.

For comparison, I list below the total number of voyages from West Florida to various places in the British empire, 1763–1778.

New York	212	Boston	4
Charleston	149	Virginia	3
Philadelphia	61	Liverpool	3
Jamaica	50	Bahamas	2
Georgia	40	Dover	1
London	28	Grenada	1
St. Augustine	17	St. Vincent	1
Rhode Island	12	Waterford	1

Appendix 2
Deerskin Prices, 1768–1778

From letter book of John Fitzpatrick (prices in sols per pound for undressed skins).

August	1768	45
September	1768	35
April	1769	40
June	1769	33
July	1769	37
August	1769	35
June	1771	48
June	1774	35
January	1775	37
June	1775	35
August	1775	40
January	1777	40
January	1778	30
September	1778	40

A sol, or sou, was a French coin equivalent to one-tenth of a bit or to one-eighteenth of a Spanish dollar, for which the standard exchange at the time was four shillings and eightpence sterling. In English money, then, Fitzpatrick's skins ranged in price from one shilling and ninepence per pound to two shillings and tenpence per pound.

Appendix 3
Agents for Military Adventurers

List of receivers of money for the Company of Military Adventurers, appointed at a general meeting of the company at Hartford on 18, 19, and 20 November 1772, as reported in the *Providence Gazette,* 30 January 1773.

Captain Hugh Ledlie of Hartford
Captain Aaron Hitchcock of Suffield
Captain Amos Walbridge of Stafford
Captain John Ellsworth of Middletown
Captain Elihu Humphrey of Simsbury
Captain Lemuel Stoughton of
  East Windsor
Captain Samuel Chapman of Tolland
Major Henry Champion of Colchester
Captain Nathaniel Loomis of Windsor
Captain Isaac Bidwell of Farrington
Major David Baldwin of Milford
Lieutenant Nathaniel Moss of
  Wallingford
Mr. Stephen Hopkins of Waterbury
Colonel Leveret Hubbard of
  New Haven
Captain Giles Russel of Stonington
Lt. William Dennison of Stonington
Lieutenant Jedediah Hide of Norwich
Mr. Nathaniel Bishop of Norwich
Mr. William Coit of New London
Captain Asa Kinnea of Preston
Captain Aaron Cleveland of Canterbury
Captain David Holmes of Woodstock
Lt. Thomas Knolton of Ashford

Major Joseph Storrs of Mansfield
Captain Nathaniel Porter of Lebanon
Ensign Stephen Richardson of Coventry
Colonel Israel Putnam of Pomfret
Colonel Ebenezer Silliman of Fairfield
Captain Thaddeus Benedict of Danbury
Major David Waterbury of Stanford
Captain Samuel Elmer of Sharon
Lieutenant Shubael Griswold of
  Torrington
Lieutenant Eleazar Curtis of Kent
Mr. Archibald McNeal, Jr., of Litchfield
John Foxcraft, Esq., of Philadelphia
Joseph Clarkson, Esq., of Boston
Mr. Clement Biddle of Philadelphia
Thomas Gray, Esq., of Boston
Peter Vanbrugh Livingston, Esq., of
  New York
Colonel James Putnam of Worcester
Capt. William Thompson of Carlisle
  in Penn.
Capt. Isaac Peabody of New Canaan
  (in Albany Co.)
Mr. Adonijah Waterman of Lenox
  (in Berkshire Co.)
Mr. John Phelps of Norfolk

Appendix 4
Slave Transactions, 1764–1779

Date	Seller and Buyer	M	B	W	G	Price/value	Name	Remarks	Source
2 Feb. 1764	Henry Fairchild (atty. for Robert Farmar & Thomas Miller) to auction?	2	—	—	—	N.g.	Peter, Prince	Part of exchange for Dauphin Is. cattle	CO5/602:437
12 Sept. 1765	Robert Harley (mariner) to Elizabeth Lutman (spinster)	5	—	4	1	£500 ($2,150)	*men:* Levant, Chester, Cudjoe, Jack, Prince; *women:* Chloe, Eve, Reho, Peggy; *girl:* Sally	All females marked R. H. on right shoulder. Sally is Chloe's daughter	CO5/602:184
21 Oct. 1765	John Hatcher (ship's carpenter) to William Shaw (Pensacola merchant)	2	—	—	—	£245.18.9 ($1,056)	Cato, Caesar	Slaves are ship's carpenters & are given as security for a debt	CO5/602:98

Date	Seller to Buyer				Notes	Slaves	Remarks	Source
17 Dec. 1765	Henry Driscoll (merchant) and Henry Lizars to Sir John Lindsay	8	—	3	1 Cumberland was lost at sea and could not be delivered. He was assessed at £56 ($241)	men: Michael, Cumberland, Geoffrey, Samuel, Fortune, Charles, Caesar, Quachiba; women: Diana, Lucy, Venice (& child)	All slaves are security for a debt of £487.13.8 ($2,097)	U.S. National Archives, General Land Office (Div. D)
9 Dec. 1766	Henry Skelton (merchant) to David Williams	1	—	—	$80	Pompey	Spots on right cheek	CO5/602:190
10 Jan. 1767	Governor Johnstone to James Primrose Thompson	—	—	1	— 1 guinea (nominal) ($4.25)	Phyllis	Later manumitted 7 Oct. 1768	CO5/605:349

Appendix 4 (Continued)

Date	Seller and Buyer	M	B	W	G	Price/value	Name	Remarks	Source
1 Apr. 1768	Henry Lizars to David Hodge, Chas. Strachan & Daniel Ward	11	—	2	2	N.g.	*men*: Bluckwal, Young Caesar, Old Caesar, Charles, Jeoffrey, Michael, Peter, Quah, Quashee, Sam, Sambo; *women*: Lucy, Venus; *girls*: Lucinda, Polly	Lizars owed $6,500 and had to realize his assets, of which these slaves were part	CO5/612:330
25 Apr. 1769	Arthur & Elinor Neil to Valens Stephen Comyn	4	2	—	2	N.g.	*men*: Anthony, Brutus, Caesar, Zama; *boys*: Charles, London; *girls*: Helen, Maria	Security for several sums paid by Jacob Blackwell, who received use of the slaves	CO5/605:257; 612:186
8 June 1769	Daniel Mortimer to David Ward	3	—	—	—	N.g.	Adam, Hercules, Windsor	Security for debt of £93.8.8 ($402)	CO5/612:57

Date	Parties				Amount	Slave names	Description	Reference
22 July 1769	Alexander Moore to Robert Smyth (of Charleston)	3	—	2	— £151.1.6 ($650)	Cyrus, Dublin, Scipio; Molly, Sukey	Security for a debt. When Molly died, Moore substituted Phyllis	CO5/612:10
31 July 1769	Valens S. Comyn to Thomas Comyn, Stephen Comyn & Nicholas Donnithorne	4	3	6	— N.g.	*men:* John, London, Richard, Thomas; *boys:* Hero, Robert, William; *women:* Elizabeth, Harriet, Juliet, Mary, Sara, Zaya	Part payment for an unspecified large debt	CO5/612:31
11 Jan. 1770	Isaac Monsanto to John Waugh (mariner of London)	1	—	1	— N.g.	Prince, Princess	Security for debt of $457	CO5/612:92
17 Feb. 1770	James Sutton to Montfort Browne	—	—	1	— N.g.	Sam	Indian slave, security for debt of £37.6.8 ($160)	CO5/612:57

Appendix 4 (Continued)

Date	Seller and Buyer	M	B	W	G	Price/value	Name	Remarks	Source
20 Mar. 1770	Pierre Rochon to his mistress & their children	—	2	1	4	Manumission gratis	*mother*: Marianne; *boys*: Alexy (5½ yrs), Charles (11); *girls*: Charlotte (9½), Claudine (4), Isabel (7½), Rose (2½)	Children to be freed at once; Marianne after Rochon's death; each child to have a yearly heifer	CO5/613:15
22 Mar. 1770	David Ross to William Dutton	—	—	1	—	$220	Statira		CO5/613:124
10 May 1770	Thomas Hammond to Elias Durnford	1	—	—	—	$180	Jack		CO5/612 (no folio no.)
2 June 1770	Louis & Jean Lusser to Elias Durnford	6	—	—	—	N.g.	Jaco, Joseph, Levere, Louis, Magroar, Scipio	Security for debt of £150 ($645)	CO5/612:96
4 June 1770	Isaac Monsanto to Messrs. McGillivray & Struthers	1	—	1	—	N.g.	Caesar, Dolly	Security for debt of $436	CO5/612:96

Date	Transaction	No.		Price	Slaves	Notes	Reference
30 Oct. 1770	Guillaume Loyson (silversmith) to Marianne, Creolle	1	—	$80 (price of manumission)	Marianne, Creolle		CO5/613:58
29 Dec. 1770	John Watkins to David Hodge, William Godley & George Raincock	1	—	N.g.	N.g.	Included in sale of house for £50 ($215)	CO5/612 (no folio no.)
23 July 1771	David Hodge to James Bruce & William Godley	1	—	N.g.	Prince, Celia	Apparently a deal to prevent seizure of Hodge's property by Phillips Comyn	CO5/612:162
24 July 1771	William Aird (carpenter) to James Aird (carpenter)	5	—	£60 ($258) £80 ($344) £55 ($237) £70 ($301) £72 ($310)	Bob—carpenter, Hector—carpenter, Nero—sawyer, Pompey—sawyer, Primus—sawyer	Slaves come with premises rented by Jas. Aird for £58 ($249) per annum	CO5/612:388

Appendix 4 (Continued)

Date	Seller and Buyer	M	B	W	G	Price/value	Name	Remarks	Source
25 Aug. 1771	Marie Marguerite Dufrene to Henry	1	—	—	—	Manumission gratis	Henry	A 27-year-old métif (mustee)	CO5/613:13
18 Sept. 1771	Guillaume Loyson to Robert Foley (shipwright)	—	1	—	—	$200	Chapelle		CO5/613:8
28 Sept. 1771	Thomas Ongston to Benjamin Ward	1	—	—	—	£95 ($407)	Louis	Mulatto	CO5/613:62
3 Nov. 1771	René Roi to freedwoman Charlotte	—	1	—	—	Free gift	Joseph	Joseph was son of Roi's slave Perrine	CO5/613:12
9 Nov. 1771	Isaac Monsanto to John Scott	2	—	2	—	N.g.	César, Prince; Dolly, Princess	Security for debt of £212.8.0 ($912)	CO5/612:210
14 Dec. 1771	Isaac Monsanto to David Hodge, William Godley & George Raincock	—	—	—	1	£90.0.8½ ($387)	Franchonet	Mulattress	CO5/612:210

Date	Transaction					Price	Slaves	Notes	Reference
1 Jan. 1772	Pierre Rochon to Alexander Mackintosh	2	—	1	—	N.g.	Antoine, Pierre; Helen	Security for a debt of $1,337.8¼ which Rochon never paid, so he surrendered these slaves formally on 5 Aug. 1773	CO5/612:349
7 Jan. 1772	Pierre Rochon to Alexander Mackintosh	2	—	2	—	N.g.	Antoine, Pierre; Helen, Madeleine	Security to Mackintosh as cosigner of a bond with Rochon	CO5/612:313
7 Jan. 1772	Pierre Rochon, Jr., to Elizabeth Benoist, widow of Pierre Rochon, Sr.	7	1	6	—	$2,175 (av. $155 each)	men & boys: Aboy, Baptiste, Basil, Charlot Jr., Dohi, Jean-Baptiste, Julien, Louis; women: Adelaide, Henriette, Iris, Juliet, Marianne, Victoire	Jean-Baptiste was the son of Henriette	CO5/613:98

Appendix 4 (Continued)

Date	Seller and Buyer	M	B	W	G	Price/value	Name	Remarks	Source
20 Jan. 1772	Isaac Monsanto to John Ritson	—	—	1	—	$570 (possibly a mistranscription for $270)	Franchonet	Mulattress; Monsanto will have her back on paying $270	CO5/612:219
21 Jan. 1772	Pierre Rochon to John Miller & Peter Swanson	1	—	1	—	N.g.	Baptiste, Madeleine	Security for debt of £125.4.8 ($539) which remained unpaid, so slaves became Miller's & Swanson's on 5 Aug. 1773	CO5/612:355
6 Feb. 1772	Joseph Maison to Lazare Estardy	—	1	1	1	$600	*woman*: Magdalene; *girl*: Genevieve; *boy*: François	Magdalene was a mulattress. Genevieve (aged 13) and François (aged 6) were her children	CO5/613:35

Date	Transaction					Price	Slaves	Notes	Reference
2 Mar. 1772	Jacob Blackwell to John Stephenson	8	—	—	2	$1,260 for all	Aberdeen, Davy, Dublin, Essex, George, Hector, Primus, Woolwich; Kitty, Phyllis		CO5/613:74
2 Mar. 1772	John Gradenigo to Patrick Morgan	—	—	—	2	£45.18.4 ($198) for both	Diana, Sarah		CO5/613:52
28 Mar. 1772	Isaac Monsanto to John Stuart	2	—	2		N.g.	César, Prince; Dolly, Princess	Security for a debt of £232.6.8 ($999)	CO5/612:259
9 Apr. 1772	Pierre Baptiste to William Ogilvie	—	1			$500	Tonette	Mulattress	CO5/613:67
27 Apr. 1772	Pierre Rochon to Messrs. McGillivray & Struthers	4	1	4		N.g.	men: Cadet, Charlot, Claude, François; boy: Nicolas; women: Genevieve, Henriette, Manon, Margot	Security for a debt of $1,230.5½ r which was never paid. The slaves became McGillivray's and Struthers's on 5 Aug. 1773	CO5/612:359

Appendix 4 (Continued)

Date	Seller and Buyer	M	B	W	G	Price/value	Name	Remarks	Source
4 May 1772	David Waugh to John Miller & Co.	4	—	—	—	£51.16.5 ($223)	Abba, Jambo, Sandy, Will	Will was Indian	CO5/612:261
9 May 1772	James Richards (mariner) to John Miller & Peter Swanson	—	—	1	1	$250 for both	Coula, Flora	Mother & child	CO5/613:97
2 June 1772	Pierre & Jean Rochon to Charles Parent	—	3	—	1	N.g.	Bogadous, Francis, Nicolas; Henriette	Security for a debt of £578.10.6 ($2,488)	CO5/612:270–78
13 June 1772	Pierre & Jean Rochon to Charles Parent	—	3	—	—	$1,000 for all three	César, Guillaume, John	John was a mulatto	CO5/613:97
5 Aug. 1772	David Waugh to David Ross	1	—	—	—	$120	Sandy	Sandy was then actually with William Sanders, a Pensacola silversmith	CO5/613:108

Date	Transaction				Price	Slave names	Notes	Reference
27 Aug. 1772	David Waugh to David Ross	1	1	—	£59.6.5 ($255)	Abba, Jambo, Will (Indian)	Waugh's sale to Miller of 4 May may have fallen through. Here he obtains a better price	CO5/613:123
23 Oct. 1772	Pierre Rochon to Valentin Doullin (a baker)	—	1	—	$200	Gabriel	Gabriel was aged 13	CO5/613:127
4 Nov. 1772	Pierre Rochon to Jean Rochon (his younger brother)	10	3	—	N.g.	Antonio, Christopher, François, Gohis, Jean, Jean-Pierre, Leveille, Pagadon, Paris, Paul; Diana, Magdalene-Teresa, Maria	The slaves, with horses and cattle, composed 25% of the estate of Pierre Rochon, Sr, valued at $2,296	CO5/613:164
31 Dec. 1772	Pierre & Jean Rochon to Daniel & Benjamin Ward	3	1	—	N.g.	Gohis, Goyau, Jean; Diana	Security for a loan of £227.15.7 ($979) which was never repaid. The slaves became the Wards' on 5 Aug. 1773	CO5/612:352; 613:154

Appendix 4 (Continued)

Date	Seller and Buyer	M	B	W	G	Price/value	Name	Remarks	Source
20 Feb. 1773	Isaac Monsanto to John Stuart	1	—	1	—	$625 for both	Prince, Princess		CO5/613:147
1 Mar. 1773	Julie de la Brosse Azemare to Messrs. Mc-Gillivray & Struthers	2	1	2	2	N.g.	*men:* Baptiste, Charles; *boy:* Michel; *women:* Madelon, Venise; *girls:* Isabel, Mary Jane	Security for a debt of $2,208. The girls were Madelon's daughters	CO5/612:360
4 May 1773	Antonio Garson to Maria	—	—	1	—	£5 ($22), nominal manumission payment	Maria	Maria was aged 27	CO5/613:152
1 July 1773	Robert Farmar to James Miller & James Mant	4	—	2	—	N.g.	Hector, Jack, Mercury, Valentin; Julie, Sarah	Security for a debt of £233.6.8 ($1,003) which was paid prior to Mant's death	CO5/612:347, 641

Date	Transaction				Price	Slaves	Notes	Reference
5 Aug. 1773	Pierre & Jean Rochon to Alexander Mackintosh	2	—	3	— $549.1½ r	Congo, François; Lachature, Rose, Therese		CO5/613:176
6 Nov. 1773	William Dutton to William Pountney	1	—	—	2 N.g.	Will; Statira, Venus	Security for a debt of £112.17.7 ($485)	CO5/612:393
7 Dec. 1773	Edmund R. Wegg to Jean-Claude Dupont	1	—	1	— $520	Jack, Lisette	Man & wife	CO5/613:276
21 Mar. 1774	Antonio Garson to Phillips Comyn	1	—	1	— N.g.	John, Marie-Belle	Part of a package including farm animals to expunge a debt of $970	CO5/613:211
19 June 1774	Thomas Bentley to Messrs. McGillivray & Struthers	1	—	1	1 N.g.	man: Collins; woman: Madeline; girl: Genevieve (dtr.)	Security for a loan of £933.6.8 ($4,013)	CO5/612:424
28 June 1774	Gilbert Hay to Messrs. McGillivray & Struthers	2	—	3	— $809.0½ r.	Jaime, Will; Isaiah, Jane, Sue	Jaime was mulatto; Isaiah was female!	CO5/613:251

Appendix 4 (Continued)

Date	Seller and Buyer	M	B	W	G	Price/value	Name	Remarks	Source
15 July 1774	Gilbert Hay to John Miller	1	—	—	—	N.g.	Dick	Collateral for a loan of £43.13.4 ($188)	CO5/612:429
22 Aug. 1774	Phillips Comyn to Maria Belle	—	—	1	—	$200 manumission price	Maria Belle		CO5/613:238
Sept. 1774	Oliver Pollock to James Rumsey	1	—	—	—	$400	N.g.	The man ran back to Pollock within 3 months	James, *Pollock*, p. 55
15 Oct. 1774	Adam Chrystie to William Marshall	4	—	1	—	Valued at $1,371 for all	Alexander, Ali, Cesar, Cuffee; Sally	Security for payment of $36 annually for hire of 6 blacks, one of whom is unnamed	CO5/612:573
24 May 1775	John Mackintosh & Co. to Jean-Claude Dupont	1	—	1	—	$600 for both	Antoine, Madeleine	Man & wife	CO5/613:350

Date	Parties				Value	Slaves	Notes	Reference
21 Aug. 1775	Samuel Carney to John Mc-Gillivray & Co.	1	—	—	— £58.6.8 ($251)	Prince		CO5/613:286
10 Oct. 1775	Richard Ellis to McGillivray, Struthers, Swanson, Mackintosh & Co.	8	— 6	— N.g.	Durham, Ganymede, George, Lawrence, London, Reuben, Trojan & [?]; Deborah, Grace, Peggy, Sarah, Sylvia & [?]	Collateral for a loan of $3,872.6½ r	CO5/612:468
10 Oct. 1775	Oliver Pollock to [?]	—	— 2	— $500 for both	N.g.		James, *Pollock*, p. 55
31 Oct. 1775	William Aird to James Aird	5	—	— £337 ($1,449) for all	Bob, Nero, Hector, Pompey, Prince		CO5/613:286
11 Nov. 1775	William Dutton to Benjamin Adams (of London)	1	—	— N.g.	Toni	Part collateral for a loan of £446.5.1 ($1,919)	CO5/612:485

Appendix 4 (Continued)

Date	Seller and Buyer	M	B	W	G	Price/value	Name	Remarks	Source
28 Nov. 1775	Stephen Watts & Samuel Flowers to David Williams	24	2		18	— N.g.	*men & boys:* Caly, Congo, Cowrie, Dimbo, Freeman, George, Harry, Ishmael, Jack, Juba, Jungo, Jupiter (twice), Marmada, Masa, Musa, Numa, Pater, Peter, Philip, Prince, Sambo, Sancho, Sandy, Will, Wilkes; *women:* Catherine, Charlotte, Cumba, Dantuma, Fama, Fary, Fatima (twice), Feruba, Mahaly, Maria, Mary-Anne, Mona, Nanny, Peggy, Sal, Sara, Sarah	Uncommonly high proportion of African names probably means "new" blacks. All are part of the collateral for a loan of £5,500 ($23,650)	CO5/612:523

Date	Transaction			Price	Name(s)	Notes	Source
27 Dec. 1775	William Canty to William Johnstone	— 1 —	N.g.		Mary	Collateral for a loan of $144. After 2 years, if interest on loan has been paid regularly, Mary is to be paid $1 a month	CO5/612:512
13 Apr. 1776	James Bruce to Cadwallader Morris	8 — 3 —	N.g.		Aberdeen, Caithness, Dick, Dublin, Glasgow, Kingston, Tom, York; Chloe, Roxana, Statira	Part security for a debt of $806.6r	CO5/612:565
30 May 1776	William Dunbar to Poupet	1 —	—	$270	N.g.	A "new" black	Roland, Dunbar, p. 24
7 June 1776	William Dunbar to [John?] Mackintosh	5 —	—	$260 each	N.g.		Roland, Dunbar, p. 24

Appendix 4 (Continued)

Date	Seller and Buyer	M	B	W	G	Price/value	Name	Remarks	Source
28 June 1776	Thomas Westcott to John Miller (of Manchac) & Anthony Summersall (of St. Kitts)	1	2		1	— N.g.	*man:* Aesop; *boys:* Boto, Cesar; *woman:* Nancy	Security for a debt of £144.13.4 ($622)	CO5/613:481
17 Oct. 1776	William Dutton to Thomas Durham	2		2		— N.g.	Bob, Cato; Lucy, Sylvia	Security for a debt of $960	CO5/612:611
19 Aug. 1777	Simon McCormack to Miller, Mackintosh & Co.	1				— $350	Scipio		CO5/613:431

Date	Parties					Price	Names	Notes	Source
19 Aug. 1777	Simon McCormack to McGillivray, Struthers, Swanson, Miller & Mackintosh	—	1	2	1	N.g.	*boy:* James; *women:* Nancy, Peggy; *girl:* Molly	Security for a debt of £148.6.11 ($637)	CO5/612:615
29 Oct. 1777	Isaac Roberts to Andrew McLean & William Clark & Co.	1	—	—	—	N.g.	Frank	Collateral for a debt of £53.7.6 ($229). Frank was a cooper	CO5/612:620
14 Dec. 1777	William Dunbar (acting for Alexander Ross) to [?]	1	—	—		260 lbs. indigo (barter)	N.g.	At Pointe Coupée	Roland, *Dunbar,* p. 56
18 Jan. 1778	William McIntosh to John Campbell and William Dunbar	4	—	—		N.g.	Congo, Peter, Quamina, Telemaque	Part security for a debt of $1,663.1r	CO5/612:644

Appendix 4 (Continued)

Date	Seller and Buyer	M	B	W	G	Price/value	Name	Remarks	Source
9 Feb. 1778	John West (carpenter) to Andrew Rainsford	1	—	—	—	N.g.	Job	Collateral for a loan of £70 ($301)	CO5/612:638
21 Apr. 1778	John Stewart [Stuart?] to John McGillivray	5	—	3	—	N.g.	Derry, England, Hannibal, Ireland, Yellow; Diana, Peggy, Jennet	Security for a debt of £250 ($1,075)	CO5/612:672
20 June 1778	Langley Bryant to John Stuart	1	—	—	—	£60 ($258)	Britain		CO5/613:486
21 Sept. 1778	Richard Ellis to John McGillivray	2	—	—	8	N.g.	Daniel, Dick; Betty, Chloe, Chovilla, Fanny, Hannah, Lucy, Sinnah, Tabby	Collateral for a short term loan repayable by 1 Oct. 1778	CO5/613:532

Date	Transaction				Value	Slaves	Remarks	Source
28 Oct. 1778	William Williams & William Ferguson to Messrs. McGillivray, Struthers, Swanson, Miller & Mackintosh	2	—	1	— N.g.	James, Joe; Delia	Part of collateral for a debt of unspecified size	CO5/613:483–86
19 Nov. 1778	Thomas James to Messrs. Morgan & Mather	6	—	3	— N.g.	Aaron, Dembo, Guy, Jack, Sambo, Simon; Farsall, Phyllis, Sollom	Part security for a loan of unspecified size	CO5/612:672
27 Jan. 1779	Edward Taylor to Arthur Strother & John Stephenson	8	—	—	— N.g.	Dimbo, Hammer, Marmie, Maulick, Muggy, Othello, Sambo, Sidney	Part collateral for a debt of $2,030	CO5/612:686
3 Mar. 1779	Nehemiah Carter to Joshua Howard	1	—	—	— $400	Thom		CO5/613:522

Appendix 4 (Continued)

Date	Seller and Buyer	M	B	W	G	Price/value	Name	Remarks	Source
10 Mar. 1779	John Austin to Messrs. Struthers, Swanson, Midler & McGillivray	—	1	1	—	N.g.	Aimwell, Moll	Collateral for a $500 loan	CO5/612:695
26 June 1779	Charles Weatherford to William Clark	6	—	1	2	$1,800	*men:* Hercules, Jack, Jeffrey, London, Martin, Tom; *woman:* Elsie; *girls:* Charlotte, Delia		CO5/616:75

Note: The total number of men, boys, women, and girls in each transaction is given under the columns M, B, W, and G. "N.g." indicates that the price or name is not given in the records, and "r" abbreviates "ryal," which is one-eighth of a Spanish dollar.

Appendix 5
Imports to Britain from Florida, 1773–1775
(value in pounds sterling)

Item	1773 Quantity	1773 Value	1774 Quantity	1774 Value	1775 Quantity	1775 Value
Bear Black	166	53.19.0	29	9.8.6	60	19.10.0
Beaver	514	89.19.0	1,612	282.2.0	17	2.19.6
Cat	5	1.0	79	16.5	35	7.3
Deer in Hair	39,709	4,963.12.6	131,109	16,388.12.6	62,141	7,767.12.6
Deer India ½ Drest (lbs.)	14,055	878.8.9	51,438	3,214.17.6	85,748	5,359.5.0
Elk	26	7.3.0	137	37.13.6	141	38.15.6
Fox	—	—	60	3.15.0	—	—
Martin	—	—	13	8.8	—	—
Mink	1	1.9	113	9.17.9	12	1.1.0
Musquash	—	—	4	2.0	—	—
Otter	267	46.14.6	539	94.6.6	477	83.9.6
Panther	1	1.0.0	12	12.0.0	—	—
Raccoon	—	—	30	15.0	36	18.0
Tyger	—	—	8	4.0.0	—	—
Wolf	4	1.0.0	1	5.0	—	—
Total	—	6,041.19.6	—	20,059.0.4	—	13,273.18.3

Note: Original spellings have been retained. Dashes indicate that no imports were recorded.
Source: T64/276, Public Record Office, Kew, Great Britain.

Notes

Introduction

1.   Arthur B. Keith, ed., *Selected Speeches and Documents on British Colonial Policy, 1763–1917* (London: Oxford University Press, 1933), 1:4.
2.   William L. Grant and James Munro, eds., *Acts of the Privy Council of England*, Colonial Series (London: Lords Commissioners of the Treasury, 1908–1912), 4:668.
3.   Hillsborough to Gage, 31 July 1770, Great Britain, Public Record Office, Kew, CO 5/88:100.
4.   *South Carolina Gazette*, 3 October 1769.
5.   "Letters from a Pennsylvania Farmer to the Inhabitants of the British Colonies" Letter 8 reproduced in ibid., 29 February 1768.
6.   Grant to Hillsborough, 19 October 1770, K. G. Davies, ed., *Documents of the American Revolution, 1770–1783* (Shannon: Irish University Press, 1972–1981), 2:216.
7.   Cecil Johnson, *British West Florida, 1763–1783* (New Haven: Yale University Press, 1942), and Clinton N. Howard, *The British Development of West Florida, 1763–1769* (Berkeley: University of California Press, 1947).

1: Immigration

1.   *London Gazette*, 14 November 1763; *New York Gazette*, 4 March 1765.
2.   Great Britain, Public Record Office, Kew, PRO 30/47 (Egremont MSS) 14:88–89.

3. Robin Fabel, "George Johnstone and the 'Thoughts concerning Florida'—A Case of Lobbying?" *Alabama Review* 19 (July 1976): 163–76.

4. *New York Gazette*, 4 March 1765.

5. Keith, *Selected Speeches*, 1:7–8.

6. Howard, *British Development*, pp. 65, 80.

7. Frederick A. Pottle, ed., *Boswell's London Journal, 1762–1763* (New Haven: Yale University Press, 1950), pp. 124–25.

8. Howard, *British Development*, p. 80.

9. CO5/634:209.

10. Howard, *British Development*, p. 87; CO5/607:216–19.

11. *Scots Magazine*, February 1765; *South Carolina Gazette*, 14 December 1765; *Georgia Gazette*, 10 January 1765; *New York Gazette*, 25 February 1765.

12. Johnstone to Pownall, 2 April 1766, CO5/574:968.

13. *Massachusetts and Boston Newsletter*, 9, 30 August 1764.

14. *Georgia Gazette*, 10 May 1764.

15. Ibid., 10 May 1764, 20 June 1765.

16. *Scots Magazine*, May 1766.

17. *Maryland Gazette*, 9 May 1765; *Caledonian Mercury*, 3 August 1763.

18. *New York Journal*, 12 February 1767.

19. *South Carolina Gazette*, 22 October 1764.

20. *Maryland Gazette*, 27 June 1765.

21. *South Carolina Gazette*, 24 August 1765.

22. CO5/602:82.

23. CO5/603:2.

24. Howard, *British Development*, p. 87.

25. Great Britain, Public Record Office, Kew, AO1/1261, and see also my "Letters of 'R'," *Louisiana History* 24 (Fall 1983).

26. Elias Durnford to Hillsborough, 5 June 1768, CO5/69:443.

27. J. Barton Starr, "Campbell Town: French Huguenots in British West Florida," *Florida Historical Quarterly* 54 (April 1976): 532–47.

28. *New York Mercury*, 3 March 1766.

29. *New York Gazette*, 27 July 1767.

30. Anon., Charleston, 12 December 1766, Clifford K. Shipton, comp., *Early American Imprints, 1639–1800*, #10252.

31. Starr, "Campbell Town," p. 534.

32. *Boston Chronicle*, 18 April 1768.

33. Johnstone to Halifax, 19 February 1765, Dunbar Rowland, ed., *Mississippi Provincial Archives, 1763–1766: English Dominion* (Nashville: Brandon Printing, 1911), 1:255, hereinafter cited as *MPAED*.

34. Browne to Hillsborough, 6 July 1768, CO5/585:150.

35. *Boston Chronicle*, 1 August 1768.

36. Mark Van Doren, ed., *Travels of William Bartram* (Philadelphia, 1791; reprint, New York: Dover Publications, 1955), p. 348.

37. *Massachusetts Gazette and Boston Newsletter*, 24 October 1765.

38. John Stephenson to Montfort Browne, 15 February 1774, Dartmouth MSS, Staffordshire County Record Office, D(W) 1778/I/ii:949.

39. CO5/634:141.

40. CO5/630, Council Minutes for 19 April 1773.

41. Bernard Romans, *A Concise Natural History of East and West Florida* (New York, 1775; reprint, Gainesville: University of Florida Press, 1962), pp. 191–98.

42. Robert R. Rea and Milo B. Howard, Jr., comps., *The Minutes, Journals, and Acts of the General Assembly of British West Florida* (University, Ala.: University of Alabama Press, 1979), p. 360.

43. James Robertson to Jeffrey Amherst, 8 March 1764, CO5/83:137.

44. *South Carolina Gazette,* 14 January 1764.

45. *London Magazine,* February 1766.

46. Lawrence H. Gipson, *The British Empire before the American Revolution* (New York: Knopf, 1958–1970), 9:213.

47. Johnstone to Pownall, 1 April 1766, CO5/583:433.

48. Elias Durnford, "Description of West Florida," 15 January 1774, CO5/591:24.

49. Roy A. Rauschenberg, "John Ellis, Royal Agent for West Florida," *Florida Historical Quarterly* 62 (July 1983): 12.

50. J. Barton Starr, *Tories, Dons, and Rebels: The American Revolution in British West Florida* (Gainesville: University Presses of Florida, 1976).

51. Johnstone to Pownall, 1 April 1766, CO5/583:433.

52. Johnson, *British West Florida,* p. 149.

53. Starr, *Tories,* p. 231.

54. CO5/630, 631, 634, 635 passim.

55. Starr, *Tories,* p. 211.

56. CO5/631, Council Minutes for 26 February 1776.

57. Robin Fabel, "Bernard Lintot: A Connecticut Yankee on the Mississippi, 1775–1805," *Florida Historical Quarterly* 60 (July 1981): 97.

58. Fitzpatrick to John Stephenson, 23 May 1785, Margaret F. Dalrymple, ed., *The Merchant of Manchac: The Letterbooks of John Fitzpatrick, 1768–1790* (Baton Rouge: Louisiana State University Press, 1978), p. 418. Italics mine.

59. U.S. Department of Commerce, Bureau of the Census, *Historical Statistics of the United States: Colonial Times to 1970* (Washington, D.C.: U.S. Government Printing Office, 1975), 2:1168.

2: Blacks in West Florida

1. *Georgia Gazette,* 1 December 1763.

2. Howard, *British Development,* pp. 54, 102.

3. *Georgia Gazette,* 9 February 1764.

4.   Ibid., 27 December 1764, 10 January 1765.

5.   PRO 30/47/14:89; Fabel, "Thoughts," pp. 164–76.

6.   Library of Congress, Records of the General Land Office, G Division, Council Minutes for 5 May 1769.

7.   Rea and Howard, *General Assembly*, pp. 11, 17, 18, 22, 330–36.

8.   An Act concerning Flats, Boats, and Canoes of 3 January 1767, ibid., p. 323.

9.   Both "mulatto" and "mustee" referred to the offspring of a white and a black. Here, though, the more precise meaning of "mustee," also known as "métif," is meant, namely, the offspring of a white and a quadroon.

10.   John Hope Franklin, *From Slavery to Freedom* (New York: Knopf, 1974), pp. 61–62.

11.   Rea and Howard, *General Assembly*, pp. 82, 108, 342–47.

12.   In 1767 Moore paid the poll tax on eight slaves, Aird on nineteen. Of the committee of three, only Leitch paid any poll tax and then only on one slave. CO5/577:76.

13.   W. Robert Higgins, "The Ambivalence of Freedom: Whites, Blacks, and the Coming of the American Revolution in the South," in W. Robert Higgins, ed., *The Revolutionary War in the South: Power, Conflict, and Leadership* (Durham, N.C.: Duke University Press, 1979), p. 59.

14.   Chester to Hillsborough, 25 December 1770, CO5/588:80.

15.   Jackson to the Board of Trade, 9 January, 30 May 1771, Davies, *Documents of the American Revolution*, 1:249, 337.

16.   CO5/619:94.

17.   CO5/630, Council Minutes for 4 November 1772.

18.   Library of Congress, West Florida Papers.

19.   Rea and Howard, *General Assembly*, pp. 333, 345.

20.   CO5/595:5–6.

21.   *Pennsylvania Gazette*, 9 July 1767.

22.   Stuart A. Landry, ed., Charles César Robin, *Voyage to Louisiana* (New Orleans: Pelican Printing, 1966), p. 49.

23.   *Georgia Gazette*, 21 September 1768, 20 December 1769.

24.   *South Carolina Gazette*, 20 October 1772.

25.   *South Carolina and American General Gazette*, 27 January 1775.

26.   John Fitzpatrick to Peter Swanson, 22 September 1771, 12 October 1772, to Alexander McIntosh, 28 April 1777, Dalrymple, *Merchant of Manchac*, pp. 111, 131–32, 245.

27.   Eron Rowland, ed., *Life, Letters, and Papers of William Dunbar* (Jackson, Miss.: Press of the Mississippi Historical Society, 1930), pp. 29, 30, 35, 67, 71.

28.   Villiers to d'Abbadie, 13 March 1764, Clarence W. Alvord and Clarence E. Carter, eds., *The Critical Period, 1763–1765* (Springfield, Ill.: Illinois State Historical Society, 1915), p. 228.

29.   Gordon's Journal, Johnstone to Pownall, 19 February 1765, and Aubry to the Minister, 12 February 1765, ibid., pp. 305, 435, 437.

30. For a fuller discussion of the decrees of 1766 and 1768 which sought to make the trade of New Orleans part of the closed Spanish imperial system, see John P. Moore, *Revolt in Louisiana: The Spanish Occupation, 1766–1770* (Baton Rouge: Louisiana State University Press, 1976), pp. 105–7, 113.

31. *Rivington's New York Gazetteer,* 6 July 1775.

32. Library of Congress, Records of the General Land Office, Council Minutes for 5 May 1769.

33. O'Reilly to Browne, 24 September 1769, CO5/587:42.

34. John Clark, *New Orleans, 1718–1812: An Economic History* (Baton Rouge: Louisiana State University Press, 1970), pp. 165, 176–77.

35. Fitzpatrick to Pollock, 17 October 1770, Dalrymple, *Merchant of Manchac,* pp. 20, 213; Rowland, *Dunbar,* pp. 50, 55.

36. Robert V. Haynes, *The Natchez District in the American Revolution* (Jackson, Miss.: University Press of Mississippi, 1976), pp. 69, 74.

37. Fitzpatrick to Stephenson, 23 April 1777, to O'Keefe, 18 June 1778, Dalrymple, *Merchant of Manchac,* pp. 244, 295.

38. PRO ADM/241. I am grateful to Robert R. Rea of Auburn University for this reference.

39. John W. Caughey, *Bernardo de Gálvez in Louisiana, 1776–1783* (Berkeley: University of California Press, 1934), pp. 57, 113, 121.

40. George Phyn to Sir William Johnson, 15 April 1768, Clarence W. Alvord and Clarence E. Carter, *Trade and Politics, 1767–1769* (Springfield, Ill.: Illinois State Historical Society, 1921), p. 244.

41. Extract from Edward Codrington's account book, 1773–1774, CO5/613:307–8.

42. Bill of Sale of 20 July 1775, ibid., pp. 353–54.

43. *Pennsylvania Gazette,* 23 September 1772, 14 April 1773; CO5/613:307–8.

44. CO5/613:353–54, 364.

45. CO5/612:221.

46. Morgan to Baynton and Wharton, 2, 6, 11 December 1767, February 1768, Alvord and Carter, *Trade,* pp. 126, 128, 135, 162, 331.

47. CO5/635:234, Council Minutes for 7 January 1779.

48. CO142/17:12, 17, 33, 56, 79, 121; 18:112.

49. Howard, *British Development,* pp. 89, 101.

50. Rowland, *Dunbar,* p. 10.

51. Dalrymple, *Merchant of Manchac,* p. 221, n. 12.

52. CO5/631, Council Minutes for 6 November 1776, 26 March 1777.

53. CO5/635, Council Minutes for 17 May 1779.

54. Chester to Gálvez, 28 May 1778, CO5/594:630; David Ross & Co. to John Ferguson, 11 April 1778, CO5/631, Council Minutes for 25 April 1778.

55. Johnson, *British West Florida,* p. 148.

56. CO5/631, Council Minutes for 11 October 1777.

57. Rea and Howard, *General Assembly,* p. xxv.

58. Charles Strachan to Alexander Wylly, 10 June 1767, Letter Book of Charles

Strachan, National Library of Scotland, Edinburgh, for transcriptions from which I am again grateful to Robert R. Rea.

59.   Appendix to *Collections of the Georgia Historical Society* (Savannah: Georgia Historical Society, 1913), vol. 8, after p. 259.

60.   Davies, *Documents of the American Revolution*, 1:171.

61.   Albert C. Bates, ed., *The Two Putnams: Israel and Rufus in the Havana, 1762, and in the Mississippi River Exploration, 1772–1773, with Some Account of the Company of Military Adventurers* (Hartford: Connecticut Historical Society, 1931), p. 194; Jack Sosin, *The Revolutionary Frontier, 1763–1783* (New York: Holt, Rinehart & Winston, 1967), p. 62; CO5/631, Council Minutes for 6 November 1776.

62.   CO5/631, Council Minutes for 26 February 1776.

63.   Johnson, *British West Florida*, p. 147.

64.   John Stuart to William Knox, 9 October–26 November 1778, Davies, *Documents of the American Revolution*, 15:215; McIntosh to Chester, 16 October 1778, and Chester to Germain, 27 November 1778, CO5/595:461–63, 470.

65.   Howard, *British Development*, pp. 147, 148.

66.   *South Carolina Gazette*, 29 August 1769.

67.   Grant to Hillsborough, 16 January 1770, Davies, *Documents of the American Revolution*, 2:29.

68.   *South Carolina and American General Gazette*, 5 February 1771.

69.   Elias Durnford, "Description of West Florida," 15 January 1774, CO5/591:24.

70.   Browne to the Board of Trade, 28 February 1769, CO5/577:72.

71.   Board of Trade to the King, 15 February 1769, CO5/599:173.

72.   James A. James, *Oliver Pollock* (New York: Appleton-Century, 1937), p. 54. Italics mine.

73.   CO5/585:203–4.

74.   Daniel P. Mannix, *Black Cargoes* (New York: Viking Press, 1962), pp. 169–70.

75.   Gomer Williams, *History of the Liverpool Privateers and Letters of Marque with an Account of the Liverpool Slave Trade* (London, 1897; reprint, New York: Kelley, 1966), pp. 674–75.

76.   Elizabeth Donnan, *Documents Illustrative of the History of the Slave Trade to America* (Washington, D.C.: Carnegie Institution of Washington, 1935), 2:497, 498, 537, 656.

77.   CO5/585:280–81.

78.   Rea and Howard, *General Assembly*, pp. 328, 353, 375, 393.

79.   Dalrymple, *Merchant of Manchac*, pp. 430, 431.

80.   CO5/605:349.

81.   Gage to the Secretary of War, 17 January 1767, PRO T1/458:85.

82.   PRO T1/440:114.

83.   Henry Stuart to Henry Smith and Donald McPherson, 10 May 1781, CO5/597:740.

84.   J. Leitch Wright, Jr., "Blacks in British East Florida," *Florida Historical Quar-*

terly 54 (April 1976):434.

85.   James Campbell to Elias Durnford, 19 February 1780, CO5/597:437.

86.   Petition of Peter Swanson, John Miller, William Struthers, James Mc-Gillivray, 1798, PRO 30/8/344:185–88.

87.   CO5/635, Council Minutes for 3 March 1780.

88.   CO5/616:133–35.

89.   Library of Congress, West Florida Papers, Diary of Robert Farmar, entries for 30 March and 17 April 1781.

90.   CO5/597:638.

91.   Charles to John Stuart, 26 December 1770, quoted in Davies, *Documents of the American Revolution,* 2:303.

92.   Arthur P. Whitaker, "Alexander McGillivray, 1789–1793," *North Carolina Historical Review* 5 (1928):182.

93.   Discussion of all these authors' arguments may conveniently be found in Roger Anstey, *The Atlantic Slave Trade and British Abolition, 1760–1810* (London: Macmillan, 1975), chap. 4.

94.   *Gentleman's Magazine* 42 (1772):63.

95.   Clarence W. Alvord and Clarence E. Carter, *The New Régime, 1765–1767* (Springfield, Ill.: Illinois State Historical Society, 1916), p. 307.

96.   Romans, *Concise History,* pp. 102–11.

97.   Dalrymple, *Merchant of Manchac,* p. 432.

98.   I am grateful to Roy Rauschenberg of Ohio University for allowing me to use his transcription of Gordon's undated letter from the Ellis MSS in the Linnaeus Society of London.

99.   CO5/615:128–35.

100.   CO5/602:98, 605:75–76; 612:620.

101.   U.S. Commerce Dept., *Historical Statistics,* 2:1176.

102.   Rea and Howard, *General Assembly,* pp. 327–30.

103.   CO5/577:70, 76.

104.   I am indebted again to Robert R. Rea for transcriptions from PRO Cust 16/1 (Imports and Exports, America, 1768–1773).

105.   U.S. National Archives, General Land Office, West Florida Papers (Division D).

106.   Charles Strachan to John Falconer, 22 May 1766, and to William Telfair, 24 October 1766, 7 January 1767, Letter Book of Charles Strachan.

107.   Strachan to Wylly, 10 June 1767, ibid.; Rea and Howard, *General Assembly,* p. 104.

108.   Strachan to Wylly, 4 July 1767, Letter Book of Charles Strachan.

109.   George Ross to James Bruce, n.d., PRO 30/8/343:338.

110.   Romans, *Concise History,* p. 194.

111.   M. Le Page du Pratz, *The History of Louisiana* (London, 1774; reprint, Baton Rouge: Louisiana State University Press, 1975), p. 378.

112.   Romans, *Concise History,* pp. 249, 255.

113.   du Pratz, *Louisiana,* p. 381.

114. Romans, *Concise History*, pp. 243–44.

115. du Pratz, *Louisiana*, p. 379.

116. *South Carolina and American General Gazette*, 4 July 1770.

117. John Fitzpatrick to Daniel Ward, 17 April 1769, Dalrymple, *Merchant of Manchac*, p. 45.

118. CO5/616:133–35.

119. National Archives, General Land Office (Division D).

120. CO5/631, Council Minutes for 2 May 1776, 16 September 1777.

121. CO5/634, Council Minutes for 26 December 1776; Library of Congress, West Florida Papers, Council Minutes for 8 June 1771; Rowland, *Dunbar*, p. 23.

122. CO5/634, Council Minutes for 26 December 1776.

123. CO5/631, Council Minutes for 11 November 1776, 1 September 1777.

124. Library of Congress, West Florida Papers, Council Minutes for 28 January 1772.

125. CO5/631, Council Minutes for 16 September 1777.

126. Howard, *British Development*, p. 96.

127. CO5/631, Council Minutes for 21 June 1776, 29 August 1777.

128. CO5/630, Council Minutes for 10 January, 5 May 1773.

129. CO5/631, Council Minutes for 6 November 1776.

130. CO5/635, Council Minutes for 17 May 1779.

131. Library of Congress, West Florida Papers, Council Minutes for 28 January 1772.

132. CO5/591:23.

133. John Stephenson to Montfort Browne, 15 February 1774, Dartmouth MSS.

3: The Indian and Caribbean Trade

1. J. Leitch Wright, Jr., *The Only Land They Knew* (New York: Free Press, 1981), p. 29.

2. James Robertson to Jeffrey Amherst, 8 March 1764, CO5/83:137.

3. Ian R. Christie and Benjamin W. Labaree, *Empire or Independence, 1760–1776* (New York: Norton, 1976), pp. 35, 36.

4. Thomas Whateley, *The Regulations Lately Made concerning the Colonies and the Taxes Imposed upon Them, Considered* (London: Printed for J. Wilkie, 1765), Goldsmith Kress Collection, Ralph Brown Draughon Library, Auburn University, Microfilm #10104.

5. Kathryn E. Holland, "The Path between the Wars: Creek Relations with the British Colonies, 1763–1774" (M.A. thesis, Auburn University, 1980), pp. 6, 7.

6. *South Carolina Gazette*, 22 October 1763.

7. Holland, "Creek Relations," p. 9.

8. *New York Gazette*, 7 November 1763.

9. Holland, "Creek Relations," pp. 13, 15.

10. David H. Corkran, *The Creek Frontier, 1540–1783* (Norman, Okla.: University of Oklahoma Press, 1967), p. 241.

11. *South Carolina Gazette,* 31 March 1764.

12. Corkran, *Creek Frontier,* p. 232.

13. *New York Mercury,* 19 November 1764.

14. Holland, "Creek Relations," p. 31.

15. CO5/68:183.

16. These problems are well summarized in Clarence E. Carter, "British Policy towards the American Indians in the South, 1763–8," *English Historical Review* 33 (1918): 37–55.

17. Keith, *Selected Speeches,* 1:11.

18. John R. Alden, *John Stuart and the Southern Colonial Frontier* (Ann Arbor: University of Michigan Press, 1944), pp. 244, 246.

19. Stuart to Shelburne, 1, 11 April 1767, CO5/68:216, 249.

20. Corkran, *Creek Frontier,* pp. 250–51.

21. CO5/511:115.

22. Bodleian MS North a. 2. fols. 7, 15–19.

23. Thomas Hutchins, *An Historical Narrative and Topographical Description of Louisiana and West Florida* (Philadelphia, 1784; reprint, Gainesville: University of Florida Press, 1968), p. 70.

24. Clark and Milligan to William Knox, 29 May 1777, William J. Morgan, ed., *Naval Documents of the American Revolution* (Washington, D.C.: U.S. Government Printing Office, 1964–), 8:877.

25. John Campbell to Germain, 13 May 1780, CO5/597:429.

26. Stuart to Shelburne, 28 July 1767, CO5/68:273.

27. Johnstone to the Board of Trade, 23 October 1766, CO5/584:279.

28. Gage to Hillsborough, 7 December 1770, Clarence E. Carter, ed., *The Correspondence of General Thomas Gage with the Secretaries of State, 1763–1775* (New Haven, 1933; reprint, New York: Archon Books, 1969), 1:286.

29. Dalrymple, *Merchant of Manchac.*

30. John Thomas to Peter Chester, 16 April 1772, Gage Papers, William Clements Library, Ann Arbor, Michigan.

31. Dalrymple, *Merchant of Manchac,* p. 30.

32. Fitzpatrick to Peter Swanson, 12 August 1775, to Stephen Hayward, 16 October 1777, ibid., pp. 195, 270.

33. Fitzpatrick to Peter Swanson, 7 June 1775, to McGillivray and Struthers, 10 August 1772, ibid., pp. 126, 129.

34. Fitzpatrick to John Stephenson, 14 January 1777, to William Strother, 16 January 1777, ibid., pp. 225, 226.

35. See Appendix 2.

36. Fitzpatrick to McGillivray and Struthers, 30 August 1770, Dalrymple, *Merchant of Manchac,* p. 94.

37. Same to the same, 21 January, 22 September 1771, 22 March 1774, ibid., pp. 99, 110, 168.

38. Fitzpatrick to John McGillivray, 23 June 1774, ibid., p. 170.

39. Fitzpatrick to McGillivray and Struthers, 21 January 1771, 13 May 1772, 6 January, 30 October 1774, ibid., pp. 100, 119, 140, 174.

40. Fitzpatrick to Thomas O'Keefe, 19 March 1777, to William Smith, 27 February 1784, ibid., pp. 238, 406.

41. Thomas to Peter Chester, 26 March 1777, Gage Papers.

42. *New York Mercury*, 17 August 1772.

43. Great Britain, Public Record Office, Kew, T64/276.

44. Debra L. Fletcher, "They Lived, They Fought: The Creek-Choctaw War, 1763–1776" (M.A. thesis, Auburn University, 1983).

45. Samuel C. Williams, ed., *Adair's History of the American Indians* (London, 1775; reprint, Johnson City, Tenn.: National Society of the Colonial Dames of America, 1910), p. 229.

46. Board of Trade to Hillsborough, 7 March 1768, CO5/69:141; 68:257.

47. Stuart to Hillsborough, 15 September 1768, CO5/69:571.

48. Corkran, *Creek Frontier*, pp. 259–63.

49. CO5/71:23.

50. Alden, *John Stuart*, pp. 315, 316.

51. Rea and Howard, *General Assembly*, pp. 379–83.

52. Alden, *John Stuart*, p. 319.

53. William Stiell to Peter Chester, 1 November 1776, CO5/631, Council Minutes for 5 November 1776.

54. Corkran, *Creek Frontier*, pp. 301–2.

55. Gage to Hillsborough, 5 May 1771, *Carter Correspondence of Gage*, 1:297.

56. Chester to Dartmouth, 20 November 1775, CO5/592:185–95.

57. Robert R. Rea, "Pensacola under the British, 1763–1781," in James R. McGovern, ed., *Colonial Pensacola* (Pensacola: Pensacola Series Commemorating the American Revolution Bicentennial, 1972), p. 57.

58. *Maryland Gazette*, 15 May 1777.

59. *New York Journal*, 7 July 1777.

60. CO5/595:13.

61. *New York Journal*, 12 December 1778.

62. CO5/594:343, 348.

63. National Archives, General Land Office (Division D), Council Minutes for 6 June 1778.

64. Campbell to Germain, 12 February 1780, CO5/597:329.

65. CO5/631, Council Minutes for 29 July, 13 August 1776.

66. Ibid., Council Minutes for 29 September 1777.

67. Dalrymple, *Merchant of Manchac*, p. 221, n. 12; CO5/594:12.

68. John D. Ware and Robert R. Rea, *George Gauld, Surveyor and Cartographer of the Gulf Coast* (Gainesville: University Presses of Florida, 1982), p. 178.

69. *Maryland Gazette,* 8 May, 19 June 1777.

70. *Louisiana Historical Quarterly* 8 (1925):737–38.

71. CO5/631, Council Minutes for 20 March, 6 November 1777.

72. CO5/617:153.

73. Robert Ross to ?, 28 March 1778, CO5/595:295.

74. Dalrymple, *Merchant of Manchac,* p. 276, n. 31, and Fitzpatrick to Davis, 2 April 1780, ibid., p. 345.

75. David Ross to John Ferguson, 11 April 1778, CO5/631, Council Minutes for 25 April 1778.

76. CO5/616:59.

77. Basil Keith to Peter Chester, 18 November 1776, CO5/634:441.

78. See Donna J. Spindel, "The Stamp Act Crisis in the British West Indies," *Journal of American Studies* 11 (1977):203–21, for an analysis of why the resentment that was certainly felt in the islands toward the act caused peaceful resistance, as a result of which stamp duty was collected in only a quarter of them.

79. Library of Congress, Records of the General Land Office, fols. 322–33.

80. Rowland, *Dunbar,* p. 27.

81. CO5/635, Council Minutes for 17 May 1779.

82. CO5/613:510.

83. CO5/635, Council Minutes for 17 May 1779.

84. CO5/612:550–65.

85. Ibid., fols. 624, 672, 587–89.

86. CO5/634:444.

87. CO5/631, Council Minutes for 26 December 1776, 22 January 1778.

88. Fitzpatrick to William Walker, 12 June 1777, Dalrymple, *Merchant of Manchac,* p. 253.

89. Esteban Miró to Juan Manuel de Cagigal, 8 May 1782, Stanley Arthur, ed., *Despatches of the Spanish Governors of Louisiana* (New Orleans: Works Progress Administration, 1937–1938), 6:48–49.

90. CO5/631, Council Minutes for 1 January 1778.

91. National Archives, General Land Office (Division D).

92. CO5/612:611.

93. National Archives, General Land Office (Division D).

94. CO5/631, Council Minutes for 22 January 1778.

95. *Maryland Gazette,* 8 May 1777.

96. CO5/631, Council Minutes for 11 October 1777.

97. Rea and Howard, *General Assembly,* p. xxv.

98. CO5/635, Council Minutes for 13 January 1780.

99. CO5/631, Council Minutes for 26 February 1776.

100. Chester to Dartmouth, 26 January 1773, CO5/590:136.

101. National Archives, General Land Office (Division D).

102. Chester to Hillsborough, 6 October 1772, Dartmouth to Chester, 3 March 1773, Chester to Dartmouth, 27 August 1773, CO5/590:72, 115, 235.

103. National Archives, General Office (Division D).

104. Library of Congress, West Florida Papers, Council Minutes, 24, 25 June 1771.

105. Manuel I. Pérez-Alonso, "War Mission in the Caribbean: The Diary of San Francisco de Saavedra (1780–83)" (Ph.D. diss., Georgetown University, 1953), pp. 127, 181.

106. CO5/613:417.

4: Trading with the Spanish Empire

1. *Georgia Gazette,* 7 July 1763.
2. 3 Geo. 3, c. 22.
3. *New York Gazette,* 10 October 1763.
4. *Pennsylvania Gazette,* 3 January 1765.
5. Johnstone to Pownall, 19 February 1765, PRO T1/437:224–25.
6. Beaufaine was a member of the Royal Society and of the South Carolina provincial council. From 1762 until his death in 1766, he was customs collector at Charleston.
7. T1/437:220–21.
8. *Providence Gazette,* 15 October 1767.
9. CO5/583:585.
10. *Pennsylvania Gazette,* 11 July 1765.
11. *Maryland Gazette,* 1 August 1765.
12. Halifax to the Treasury, 8 February 1765, T1/441:3.
13. Frances Armytage, *The Free Port System in the British West Indies* (London: Longmans, Green, 1953), p. 26.
14. *Georgia Gazette,* 1 August 1765.
15. Burnaby had given the Baymen a rudimentary but workable and, as time would show, durable legal code that combined arbitrary justice with elements of democracy.
16. *Georgia Gazette,* 14 December 1764.
17. Armytage, *Free Port System,* p. 38.
18. *South Carolina Gazette,* 14 December 1764.
19. Johnstone to Pownall, 23 October 1766, CO5/584:279–85.
20. *Pennsylvania Gazette,* 11 December 1766.
21. Report of Jacob Blackwell, n.d. Photostats in Florida State University Library, Tallahassee, of originals among the Shelburne Papers, William C. Clements Library, Ann Arbor, Mich.
22. CO5/613:204.
23. *Pennsylvania Gazette,* 5 March 1767.
24. *South Carolina Gazette,* 31 August 1967.
25. Ibid., 9 February, 15 June 1767; *New York Journal,* 9 July 1767, 21 January 1768.

26. Browne to Hillsborough, 7 February 1768, CO5/585:52.

27. *New York Journal*, 22 February 1768.

28. Ibid., 21 April, 28 July 1768; *South Carolina Gazette*, 4 July 1768.

29. *Providence Gazette*, 5 September 1767.

30. CO5/600:146.

31. Chester to Hillsborough, 27 September 1770, CO5/588:3. In the eighteenth century a chariot was a high four-wheeled vehicle with back seats only.

32. CO5/587:357–364.

33. Elias Durnford to Hillsborough, 12 June 1770, ibid., p. 338.

34. Library of Congress, West Florida Papers, Council Minutes for 4 February 1772.

35. Clark, *New Orleans*, p. 161.

36. Caughey, *Gálvez in Louisiana*, p. 40.

37. Gilbert C. Din, ed., *Louisiana in 1776: A Memoir of Francisco Bouligny* (New Orleans: Louisiana Collection Series, 1977), p. 12.

38. *New York Mercury*, 6 February 1764.

39. Peter J. Hamilton, *Colonial Mobile* (Boston, 1910; reprint, University, Ala.: University of Alabama Press, 1976), p. 225.

40. *South Carolina Gazette*, 31 March 1764.

41. *New York Gazette*, 16 July 1764.

42. Aubry to the Minister, 12 February 1765, Alvord and Carter, *The Critical Period*, p. 435.

43. Strachan to Petit, 7, 12 February 1765, Letter Book of Charles Strachan.

44. Johnstone to Pownall, 29 December 1765, CO5/574:951.

45. *New York Mercury*, 8 April, 18 August 1765.

46. *New York Journal*, 27 November 1766.

47. Johnstone to Pownall, 2 April 1766, CO5/574:965.

48. *New York Journal*, 19 February 1767.

49. Bates, *Two Putnams*, p. 135.

50. *New York Journal*, 4 March 1768.

51. Rea and Howard, *General Assembly*, p. 340.

52. *New York Journal*, 16 October 1766, 23 April, 20 June, 23 July, 5, 27 August, 10, 24 December 1767, 28 April, 12 May, 28 July, 13 October 1768.

53. Kip, his crew of five, and two passengers survived the wreck. A schooner from St. Augustine rescued them late in November. *South Carolina Gazette*, 14 December 1767.

54. *Louisiana Historical Quarterly* 6 (1923):712.

55. Fitzpatrick to Stephenson, 30 June 1768, Dalrymple, *Merchant of Manchac*, p. 38.

56. Antonio de Ulloa to Antonio Bucarelli, 20 July, 1, 2 August 1768, Arthur, *Despatches of the Governors*, bk. 1, vol. 1, pp. 52, 58.

57. *Louisiana Historical Quarterly* 6 (1923): 162, 519.

58. CO5/591:351.

59. *Georgia Gazette,* 21 June, 16 July 1764; *New York Journal,* 21 May 1767, 7 April, 25 August 1768.

60. Ulloa to Bucarelli, 11 November 1768, Arthur, *Despatches of the Governors,* bk. 1, vol. 1, p. 68.

61. Fitzpatrick to Stephenson, 28 September 1768, Dalrymple, *Merchant of Manchac,* p. 42.

62. Christie and Labaree, *Empire or Independence,* p. 36.

63. Clark, *New Orleans,* pp. 167–68.

64. Caughey, *Gálvez in Louisiana,* p. 46.

65. Bertram W. Korn, *The Early Jews of New Orleans* (Waltham, Mass.: American Jewish Historical Society, 1969), p. 33.

66. Dalrymple, *Merchant of Manchac,* pp. 10–76.

67. Fitzpatrick to Arthur Strother, 7 November 1769, ibid., p. 77.

68. *South Carolina Gazette,* 19 September 1769.

69. *New York Journal,* 23 November 1769.

70. Unzaga to Bucarelli, 8 July 1770, Arthur, *Despatches of the Governors,* bk. 1, vol. 2, p. 54.

71. John Thomas to Peter Chester, 26 March 1722, Gage Papers.

72. Unzaga to Bucarelli, 31 August, 11 November 1770, 22 January 1771, Arthur, *Despatches of the Governors,* bk. 1, vol. 3, pp. 3, 13, 26.

73. *New York Mercury,* 13 July 1772.

74. Francis Murphy to Barnard and Gratz, 4 March 1772, William V. Byars, *Barnard and Gratz, Merchants in Philadelphia, 1754–1798* (Jefferson City, Mo.: Hugh Stephens Printing, 1916), p. 121.

75. *Louisiana Historical Quarterly,* 11 (1928): 340.

76. CO5/613:139; 577:71. Eight years earlier a Richard Carpenter returned from Pensacola to Newport, *Pennsylvania Gazette,* 11 July 1765.

77. CO5/591:371–82, 415.

78. Unzaga to Chester, 17 July 1774, ibid., pp. 423–25.

79. I have been unable to confirm this incident, which I have found reported only in the *New York Gazette* of 16 May 1774.

80. CO5/580:311.

81. See shipping news in the colonial newspapers in general, and for more detailed information on some voyages, see CO5/613:307–8.

82. He offered $5,000 to anyone who would introduce the manufacture of saltpeter into West Florida, *Scots Magazine* 29 (January 1767): 51.

83. CO5/634, Council Minutes for 28 November 1775.

84. Chester to Germain, 21 November 1776, Headquarters papers of the British Army in America (Sir Guy Carleton Papers), Library of Colonial Williamsburg, Williamsburg, Va. 3:330 (3–6), hereinafter cited as B.H.P.

85. CO5/630, Council Minutes for 15 December 1772.

86. Fitzpatrick to William Wilton, 23 January 1777, Dalrymple, *Merchant of Manchac,* p. 230.

87.   Anonymous memorandum, n.d., B.H.P. 5:549 (1–2).

88.   Fitzpatrick to Michael Hoopock, 14 February 1777, Dalrymple, *Merchant of Manchac,* p. 236.

89.   John W. Caughey, "Bernardo de Gálvez and the English Smugglers on the Mississippi, 1777," *Hispanic American Historical Review* 12 (1932): 48–57.

90.   Log, *West Florida,* PRO ADM 51/4390. I am grateful to Robert R. Rea for this information.

91.   Gálvez to Marques de la Torre, 6 May 1777, Arthur, *Despatches of the Governors,* 7:16, and William Stiell to Sir William Howe, 3 June 1777, B.H.P. 5:561 (2).

92.   Caughey, "Smugglers," p. 57.

93.   Din, *Louisiana in 1776,* p. 62.

94.   CO5/635, Council Minutes for 17 November 1778.

95.   Gálvez to Alexander Dickson and John Stephenson, 23 August 1777, CO5/631, Council Minutes for 29 September 1777.

96.   Anonymous, n.d., B.H.P. 5:549 (3).

97.   Dickson and Stephenson to Gálvez, 17 August 1777, CO5/631, Council Minutes for 29 September 1777.

98.   Gálvez to Diego Joseph Navarro, 9 November 1778, Arthur, *Despatches of the Governors,* 8:64.

99.   Henry Stuart to John Stuart, 11 August 1777, CO5/594:135.

100.   Gálvez to Thomas Lloyd, 11 May 1777, B.H.P. 5:523 (1–40).

101.   Pickles's luck changed soon afterward. He was turned loose in a small boat by the mutinous crew of his schooner *Bostonian.* He safely reached Cuba and returned to New Orleans. Gálvez to Navarro, 12 April 1778, Arthur, *Despatches of the Governors,* 7:70.

102.   Fitzpatrick to Alexander McIntosh, 18 April 1777, Dalrymple, *Merchant of Manchac,* p. 245.

103.   CO5/631, Council Minutes for 29 September 1777.

104.   Gálvez to Lloyd, 11 May 1777, B.H.P. 5:523 (1–4), 549 (2).

105.   Memorial of Martin Navarro of 1781(?), Mississippi Provincial Archives, Spanish Dominion, Jackson, Miss., 1:401–3, hereinafter cited as MPASD.

106.   Extract from an anonymous letter from Pensacola, 10 October 1769, in *New York Journal,* 23 November, 21 December 1769.

107.   *Louisiana Historical Quarterly* 8 (1925): 731, 734.

108.   CO5/635, Council Minutes for 17 November 1778.

109.   Morgan and Ross to Ferguson, 26 March 1778, ibid., and "Loyal Subjects Residing in Louisiana" to Ferguson, 27 March 1778, CO5/631, Council Minutes for 25 April 1778. Their names were Richard Bradley, Donald Campbell, John Campbell, John Davies, William Dunbar, William Garden, James Mather, William McIntosh, Philip Moore, Alexander Ross, David Ross, George Ross, William Swanson, and David Williams.

110.   Chester to Germain, 30 May 1778, CO5/594:613.

111.   Caughey, *Gálvez in Louisiana,* p. 111.

112. "A general diary of happenings in the province of Louisiana," MPASD, 1:143.

113. Ross to Ferguson, 11 April 1778, CO5/635:165–68, Council Minutes for 25 April 1778.

114. Caughey, *Gálvez in Louisiana,* p. 76.

115. *Louisiana Historical Quarterly* 8 (1925):737; 13 (1930): 529.

116. *Rivington's Royal Gazette,* 20 March 1779.

117. Howard, *British Development,* pp. 58, 72.

118. Rea and Howard, *General Assembly,* p. xxiii.

119. Fitzpatrick to Robert Ross, 20 September 1768, Dalrymple, *Merchant of Manchac,* pp. 41–42.

120. Jack D. Holmes, "Robert Ross' Plan for an English Invasion of Louisiana in 1782," *Louisiana History* 5 (1964): 171.

121. CO5/612:32.

122. Chester to Dartmouth, 20 December 1777, CO5/591:183.

123. CO5/613:80.

124. CO5/631, Council Minutes for 22 January 1778.

125. Fitzpatrick to Ross, 20 September 1768, Dalrymple, *Merchant of Manchac,* p. 276.

126. The Memorial of Robert Ross and John Campbell to Governor Chester, 9 September 1778, *MPAED,* 8:26.

127. CO5/613:515.

128. *Louisiana Historical Quarterly,* 12 (1929):513–14.

129. Holmes, "Robert Ross," p. 162.

130. *Rivington's Royal Gazette,* 5 May 1779.

131. Memorial of Ross and Campbell, *MPAED,* 8:33.

132. Haynes, *Natchez District,* pp. 91–92.

133. CO5/635, Council Minutes for 17 September 1778.

134. Chester to Germain, 30 December 1778, *MPAED,* 9:308.

135. Fitzpatrick to Stephenson, 10 January 1782, Dalrymple, *Merchant of Manchac,* p. 391.

136. Fitzpatrick to Walker, 12 June 1777, ibid., p. 253.

137. *Louisiana Historical Quarterly* 8 (1925):738–40.

138. Petition of Bernard Lintot to the General Assembly of Connecticut, 16 January 1784, Connecticut Archives Manuscript Division, Hartford, Conn., Revolutionary War, 1763–1789, series 1, vol. 27, fol. 64.

139. Fitzpatrick to Evan and James Jones, 13 May 1772, Dalrymple, *Merchant of Manchac,* p. 121.

140. Howard, *British Development,* p. 73.

141. CO5/577:76.

142. Howard, *British Development,* pp. 74, 84, 93.

143. CO5/591:172, 183, 285.

144. *Louisiana Historical Quarterly* 6 (1923):161.

145. CO5/588:359.

146. Fitzpatrick to Evan and James Jones, 13 May, 19 July, 16 October 1772, 8,

13 May, 7 July 1773, Dalrymple, *Merchant of Manchac,* pp. 121, 124–25, 132, 149, 154, 194.

147. Johnson, *British West Florida,* pp. 72–73.

148. CO5/613:196.

149. W.P.A. Calendar of Documents in Louisiana State Archives, no. 1669.

150. CO5/613:196.

151. Francisco Collell to Bernardo de Gálvez, 16 October 1779, MPASD, 1:276.

152. CO5/635, Council Minutes for 17 May 1779.

153. Fitzpatrick to Davis, 16 May, 12 July 1779, 4 July, 28 September, 25 November 1780, 1 January 1781, 28 March 1785, Dalrymple, *Merchant of Manchac,* pp. 320, 330, 345, 357, 370–71, 373–74, 414.

154. CO5/635, Council Minutes for 16 November 1778.

155. Fitzpatrick to Shakespear, 2 March 1780, Dalrymple, *Merchant of Manchac,* p. 342.

156. Din, *Louisiana in 1776,* p. 63.

157. CO5/635, Council Minutes for 17 May 1779.

158. Judicial Records of the Spanish Cabildo, New Orleans, nos. 1781999901, 1782041501.

159. Alvord and Carter, *Trade,* p. 181.

160. CO5/591:183.

161. CO5/604:294.

162. CO5/612:31.

163. Howard, *British Development,* p. 85.

164. Rea and Howard, *General Assembly,* pp. xxii, xxiv, 128.

165. Fitzpatrick to Valens Comyn, 18 July, 1, 11 August, 1 September 1769, Dalrymple, *Merchant of Manchac,* pp. 60, 62, 67, 70.

166. W.P.A. Calendar of Documents, no. 658.

167. Howard, *British Development,* pp. 64, 74, 79, 90, 92.

168. Library of Congress, West Florida Papers, Council Minutes for 4 February 1772.

169. Rea and Howard, *General Assembly,* p. 174.

170. Dalrymple, *Merchant of Manchac,* p. 41 n. 17, p. 217.

171. Fitzpatrick to David Ross, 1 December 1775, 25 November, 10, 29 December 1776, 23 January, 13, 21 February, 21 May, 7 June, 3 July, 14 August 1777, 27 June, 7 August 1778, 20 April 1782, ibid., pp. 196, 216–17, 221, 224, 229, 235, 237, 250, 252, 255, 259, 297, 301, 402, 491. Cottonade was a coarse cotton fabric of inferior quality.

172. Howard, *British Development,* pp. 60, 74.

173. CO5/613:34.

174. Howard, *British Development,* p. 101.

175. CO5/613:34.

176. CO5/614:284.

177. CO5/634:134.

178. CO5/612:647–51.

179.   CO5/613:454–56, 497.

180.   *Louisiana Historical Quarterly* 13 (1930):521–28.

181.   Judicial Records, no. 1784080501.

182.   Fitzpatrick to Alexander McIntosh, 28 April 1777, to Patrick Morgan, 18 February 1778, Dalrymple, *Merchant of Manchac,* pp. 245, 286.

183.   *Louisiana Historical Quarterly* 21 (1938): 688.

5: Plantation Life

1.   J. F. H. Claiborne, *Mississippi as a Province, Territory, and State* (Jackson, Miss., 1880; reprint, Baton Rouge: Louisiana State University Press, 1964), 1:115.

2.   Robert R. Rea, "Planters and Plantations in British West Florida," *Alabama Review* 29 (July 1976): 219–35, 224.

3.   Samuel Wilson, Jr., "Architecture in Eighteenth-Century Florida," in Samuel Proctor, ed., *Eighteenth-Century Florida and Its Borderlands* (Gainesville: University Presses of Florida, 1975), pp. 102–39.

4.   CO5/612:142–44.

5.   *Pennsylvania Gazette,* 29 November 1764.

6.   CO5/613:422.

7.   John Stephenson to Montfort Browne, 18 February 1772, T1/494:63.

8.   Lewis C. Gray, *History of Agriculture in the Southern United States* (New York: Smith, 1941), 1:149.

9.   Rea, "Planters," p. 227.

10.   *New York Journal,* 2 March 1769.

11.   Rea, "Planters," pp. 229, 232.

12.   Daniel Hickey to Montfort Browne, 7 April 1771, T1/494:57, 59.

13.   National Archives, General Land Office (Division D).

14.   Rea, "Planters," pp. 230–32.

15.   *New York Mercury,* 16 April 1764.

16.   The following details of making and growing indigo come mainly from du Pratz, *Louisiana,* pp. 215–16.

17.   *South Carolina and American General Gazette,* 28 March 1771, 5, 12 August 1774.

18.   Jacob Blackwell to Customs Commissioners, Manchac, 25 March 1774, T1/505:308.

19.   Bodleian MS North a. 2, fols. 5, 7, 15.

20.   Robert R. Rea, "British West Florida Trade and Commerce in the Customs Records," *Alabama Review* 37 (1984): 154.

21.   James Grant to Hillsborough, 3 March 1772, CO5/552:211.

22.   *South Carolina and American General Gazette,* 22 July 1771.

23.   Fitzpatrick to David Ross, 1 December 1775, to James Craig, 16 June 1779, Dalrymple, *Merchant of Manchac,* pp. 196, 326.

24. *Rivington's New York Gazette,* 7 October 1773.

25. Fitzpatrick to Robert Ross, 20 September 1768, Dalrymple, *Merchant of Manchac,* p. 42.

26. The unit of weight here is the long hundredweight, equal to 112 pounds.

27. See the figures relating to Dunbar's slave force on 27 May 1776, in Rowland, *Dunbar,* p. 23. Rea, "Planters," p. 223.

28. Clark, *New Orleans,* p. 187.

29. Grant to Hillsborough, 16 January 1770, Davies, *Documents of the American Revolution,* 2:29.

30. Romans, *Concise History,* pp. 91–92.

31. Sir Lewis Namier and John Brooke, *The House of Commons, 1754–1790* (London: Oxford University Press, 1964), 2:607.

32. Henniker to Grey Cooper, 29 October 1770, T1/479, pt. 2, fols. 142–43.

33. Memorial of Nathaniel Green, 24 June 1771, T1/485:295.

34. CO5/613:307–8.

35. Bates, *Two Putnams,* p. 155.

36. Romans, *Concise History,* p. 115.

37. Claiborne, *Mississippi,* p. 115.

38. Bodleian MS North a. 2, fol. 7.

39. Rowland, *Dunbar,* p. 46.

40. Van Doren, *Travels of Bartram,* p. 332; Durnford to Hillsborough, 12 May 1769, CO5/587:145; Rowland, *Dunbar,* pp. 28–38.

41. Romans, *Concise History,* pp. 125–27.

42. Gray, *History of Agriculture,* 1:281–82.

43. Rowland, *Dunbar,* pp. 28–38.

44. Romans, *Concise History,* p. 127.

45. du Pratz, *Louisiana,* pp. 210–14, from which much of the following is derived.

46. Romans, *Concise History,* p. 148.

47. CO5/511:165.

48. T64/276.

49. George Ross to James Bruce, n.d., PRO 30/8/343, pt. 3, fol. 338.

50. PRO 30/8/344:219.

51. CO5/602:394; 511:86.

52. *New York Mercury,* 6 June 1766, 8 June 1772.

53. Romans, *Concise History,* p. 157.

54. CO5/632, Council Minutes for 13 October 1766; CO5/635, Council Minutes for 13 January 1780.

55. Gray, *History of Agriculture,* 1:159.

56. Rowland, *Dunbar,* pp. 47–49.

57. CO5/1228:104; 511:83, 143.

58. PRO Cust.16/1, statistics for 1770, 1771, 1772.

59. Cust. 16/1, 17/2, 17/3, 17/4, 17/5, 17/6. I am in debt to Robert Rea for notes on these customs records.

60. *Providence Gazette,* 5, 12 December 1767.
61. Landry, *Voyage to Louisiana,* p. 29.
62. Ware and Rea, *Gauld,* p. 123.
63. Elias Durnford, "Description of West Florida, 15 January 1774," CO5/591:23.
64. Bridgen & Walker and Hindley & Needham to Hillsborough, 27 March 1770, CO5/571:175.
65. Chester to Hillsborough, 24 August 1771, CO5/588:30.
66. CO5/571:407–39.
67. Landry, *Voyage to Louisiana,* p. 29.
68. Van Doren, *Travels of Bartram,* p. 334.
69. CO5/612:270.
70. Rea, "Planters," p. 231.
71. CO142/18:79.
72. Rea, "Trade and Commerce," p. 154.
73. Johnstone to Pownall, 2 April 1766, CO5/574:965.
74. Rea, "Trade and Commerce," p. 157.
75. Romans, *Concise History,* p. 141.
76. Rowland, *Dunbar,* p. 74.
77. CO5/588:595.
78. Daniel Hickey to Montfort Browne, 7 April 1771, T1/494:57.
79. Fabel, "Bernard Lintot," pp. 90–97.

6: The Maritime Life of the Province

1. CO5/613:386.
2. *New York Gazette,* 24 May 1773.
3. Great Britain, Public Record Office, Kew, Amherst Papers, W.O. 34/55:151–52. I am grateful to Robert Rea for this reference.
4. William Smith, *History of the Post Office in British North America* (Cambridge, 1920; reprint, New York: Octagon Books, 1973), p. 34.
5. Peter Kemp, ed., *The Oxford Companion to Ships and the Sea* (London: Lane, 1976), p. 625.
6. *New York Gazette,* 6 April 1764.
7. *Georgia Gazette,* 28 February 1765.
8. Smith, *Post Office,* p. 36.
9. *New York Gazette,* 13 August 1764.
10. *Georgia Gazette,* 4 April 1765.
11. *South Carolina Gazette,* 27 October 1766.
12. *New York Journal,* 23 March 1768.
13. Smith, *Post Office,* p. 36.

14. *New York Gazette,* 2 April 1764.

15. Johnstone to Pownall, 29 December 1765, CO5/574:951.

16. Report of Jacob Blackwell on West Florida, n.d., Shelburne Papers.

17. *South Carolina Gazette,* 23 November 1767.

18. *Georgia Gazette,* 14 August 1768.

19. Browne to Hillsborough, 14 February 1769, CO5/620:151.

20. From notes on Library of Congress transcripts of PRO 5/114:284 and 293–94, kindly lent by Robert R. Rea.

21. Postmaster General to Hillsborough, 7 August 1771, Davies, *Documents of the American Revolution,* 1:384.

22. Hillsborough to the governors of Jamaica, West Florida, and South Carolina, 16 November 1771, ibid., 1:426.

23. Chester to Hillsborough, 11 July 1772, CO5/589:296.

24. CO5/631:364.

25. CO5/613:374.

26. Chester to Dartmouth, 25 September 1774, CO5/592:25.

27. Dartmouth to Chester, 1 February 1775, ibid., p. 31.

28. Postmaster General to Dartmouth, 21 January 1775, Davies, *Documents of the American Revolution,* 8:253.

29. CO5/593:253.

30. Chester to Germain, 7 May 1778, CO5/594:499.

31. *Jamaica Mercury,* 24 July 1779, 24 June 1780.

32. William Stiell to Germain, 15 October 1778, CO5/595:23.

33. Chester to Germain, 30 December 1778, 17 August 1779; Germain to Chester, 2 June 1779, ibid., ff. 657, 669, 817.

34. Campbell to Germain, 10 May 1779, CO5/597:137.

35. Chester to Germain, 17 August 1779, CO5/595:817.

36. Campbell to Germain, n.d., CO5/598:15.

37. Chester to Germain, 1 April, 17 August 1779, CO5/595:677, 807, 818.

38. Ibid., p. 844.

39. Campbell to Germain, 15 December 1779, CO5/597:241.

40. Chester to Germain, 10 December 1779, CO5/595:898.

41. Germain to Campbell, 4 April 1780, CO5/597:305.

42. Campbell to Germain, 15 December 1779, Germain to Campbell, 4 April 1780, Campbell to Germain, 15 February 1781, ibid., ff. 250, 306, 585.

43. Chester to Germain, 30 October 1780, Germain to Chester, 7 March 1781, CO5/596:8, 11.

44. The *Diligence* remained there and was captured by the Spaniards on 8 April. James A. Servies, ed., Robert R. Rea, intro., *The Log of H.M.S. Mentor, 1780–1781* (Pensacola: University Presses of Florida, 1982), p. 192.

45. Chester to Germain, 18 February 1781, CO5/596:15, 23.

46. *New York Journal,* 5 February 1767.

47. *New York Gazette,* 15 November 1773.

48. *Rivington's Royal Gazette,* 20 March 1779.
49. Ware and Rea, *Gauld,* p. 226.
50. *South Carolina Gazette,* 12 November, 10 December 1763, 14 January 1764.
51. *Georgia Gazette,* 26 July 1764.
52. *New York Mercury,* 4 February 1765.
53. *South Carolina Gazette,* 19 January 1765.
54. *Georgia Gazette,* 7 March 1765.
55. *New York Mercury,* 6 May 1765.
56. *Pennsylvania Gazette,* 8 May 1766.
57. *New York Mercury,* 10 March 1766.
58. *South Carolina Gazette,* 27 October 1766.
59. *Pennsylvania Gazette,* 13, 20 November 1766, 26 March, 30 April 1767.
60. *South Carolina Gazette,* 16 February 1767.
61. *Pennsylvania Gazette,* 24 April 1767.
62. *South Carolina Gazette,* 27 July 1767.
63. *New York Gazette,* 5 October 1767.
64. *New York Journal,* 16 January 1768.
65. *South Carolina Gazette,* 18 April 1768.
66. *New York Journal,* 5 May 1768.
67. *South Carolina Gazette,* 15 August 1768.
68. *New York Journal,* 29 December 1768.
69. Library of Congress, West Florida Papers, Council Minutes for 28 February 1770.
70. *Georgia Gazette,* 3 May 1769.
71. Chester to Hillsborough, 27 September 1770, CO5/588:31.
72. *New York Journal,* 29 November 1770.
73. *South Carolina and American General Gazette,* 13 November 1770.
74. *New York Mercury,* 22 July 1771.
75. *New York Gazette,* 22 November 1772, 10 May 1773.
76. *New York Journal,* 11 October 1773.
77. *New York Gazette,* 21 February 1774.
78. *South Carolina and American General Gazette,* 6 May 1774.
79. Library of Congress, West Florida Papers, Account Book E.
80. *New York Gazette,* 6 November 1775.
81. Henry Laurens to Magnus Watson, 15 February 1770, to John Miller, 25 October 1770, to Richard Grubb, 27 October 1770, and to Felix Warley, 3 November 1770, George C. Rogers, Jr., David R. Chesnutt, and Peggy J. Clark, eds., *The Papers of Henry Laurens* (Columbia: University of South Carolina Press, 1977), 7:231–34, 384–92, 397–98.
82. CO5/588:57–61.
83. Cambel's name appears in some official documents as "Campbell," but never on anything signed by him.
84. CO5/591:351–53.

85. Johnson, *British West Florida,* p. 23.

86. *Pennsylvania Gazette,* 12 August 1772.

87. *New York Gazette,* 17 August 1772.

88. Carl Ubbelohde, *The Vice Admiralty Courts and the American Revolution* (Chapel Hill: University of North Carolina Press, 1960), pp. 15–17.

89. *MPAED,* 9:515–30.

90. Chester to Germain, 22 February 1781, CO5/596:31–32.

91. See Appendix 1.

92. Germain to Chester, 6 November 1776, CO5/619:148.

93. CO5/593:105.

94. CO5/597:137.

95. CO5/595:878.

96. Campbell to Germain, 13 May, 22 September 1780, CO5/597:425, 496.

97. *Charleston Royal Gazette,* 21 March 1781.

98. *Jamaica Mercury,* 24 July 1779, 29 April 1780.

99. Petition of Proprietors, Settlers, Merchants, etc., to Lord George Germain, 16 February 1779, CO5/580:113.

100. Rea and Howard, *General Assembly,* pp. 321, 326, 355; CO5/588:225, 370–72.

101. Oliver M. Dickerson, *The Navigation Acts and the American Revolution* (Philadelphia, 1951; reprint, New York: Barnes, 1963), pp. 203–4, 210.

102. T1/482:244.

103. Chester to Treasury, 14 March 1771, in answer to John Robinson to Chester, 7 November 1770, CO5/588:147–51.

104. T1/471, pt. 8.

105. T1/505:158, 162.

106. CO5/552:153.

107. Richard B. Sheridan, *Sugar and Slavery: An Economic History of the British West Indies, 1623–1775* (Baltimore, Md.: Johns Hopkins University Press, 1973), p. 355.

108. CO5/588:155–59.

109. Chester to Germain, 19 September 1778, CO5/595:377.

110. National Archives, General Land Office (Division D), Council Minutes for 26, 28 May 1778.

111. G. R. Barnes and J. H. Owen, eds., *Sandwich Papers, 1771–1782* (London: Navy Record Society, 1933), 1:405; 3:123.

112. CO5/596:145, 171.

113. John Boddington to William Knox, 28 September, 6 November 1778, CO5/184:449, 487.

114. John Campbell to Germain, 11 January, 19 February 1781, *MPAED,* vol. 9.

115. *Charleston Royal Gazette,* 21 March, 19 May 1781.

116. Pérez-Alonso, "Diary of Saavedra," p. 244.

117. Servies and Rea, *Mentor,* p. 160. The frigate Saavedra mentioned was probably the decayed *Stork.*

7: The Company of Military Adventurers

1.   *Dictionary of American Biography,* s.v. "Lyman, Phineas."
2.   For a discussion of the controversy, see Delphina L. H. Clark, *Phineas Lyman, Connecticut's General* (Springfield, Mass.: Connecticut Valley Historical Museum, 1964), pp. 22–26.
3.   Benjamin W. Dwight, *The History of the Descendants of John Dwight of Dedham, Massachusetts* (New York: Printed by J. F. Trow and Sons, 1874), 1:121–22.
4.   Julian Boyd, ed., *Susquehannah Company Papers* (Ithaca, N.Y.: Cornell University Press, 1930–1971), 1:50, 86; 2:53. A useful historical summary of the company may be found in Shaw Livermore, *Early American Land Companies* (New York: Commonwealth Fund, 1939), pp. 82–89.
5.   *Providence Gazette,* 3 January 1773.
6.   Historical Manuscripts Commission, *Fifth Report,* pt. 1 (London: H.M.S.O., 1876), pp. 216–18.
7.   Clarence W. Alvord, *The Mississippi Valley in British Politics* (New York: Russell & Russell, 1916), 1:345.
8.   Benjamin Franklin to William Franklin, 29 September 1766, Alvord and Carter, *The New Regime,* p. 395.
9.   Keith, *Selected Speeches,* 1:7.
10.   Memorial of Major Timothy Hierlihy and others to the West Florida Council, 5 March 1774, CO5/630.
11.   Petition of Phineas Lyman to the King in Council, n.d., Alvord and Carter, *The New Regime,* p. 262.
12.   George Washington to Thomas Lewis, 17 February 1774, John C. Fitzpatrick, ed., *The Writings of George Washington* (Washington, D.C.: U.S. Government Printing Office, 1931–1944), 3:184.
13.   *New York Journal,* 9 July 1764.
14.   Bates, *Two Putnams,* p. 14.
15.   *Providence Gazette,* 3 March 1764.
16.   Details of the various distributions may be found in David Syrett, ed., *The Siege and Capture of Havana, 1762* (London: Navy Records Society, 1970), pp. 305–13.
17.   *Providence Gazette,* 2 November 1771.
18.   Clarence E. Carter, "Some Aspects of British Administration in West Florida," *Mississippi Valley Historical Review* 1 (1914–1915):368.
19.   Phineas Lyman, "Reasons for a Settlement, 1766," Alvord and Carter, *The New Regime,* pp. 265, 289.
20.   Benjamin Franklin to William Franklin, 30 September 1776, ibid., p. 395.
21.   Johnson to Eliphalet Dyer, 22 January 1768, Boyd, *Susquehannah Papers,* 3:8. William S. Johnson had been Lyman's contemporary at Yale and was Connecticut's agent in London between 1767 and 1771.
22.   Joseph Chew to Sir William Johnson, 13 July 1768, ibid., 3:28. Sir William

worked with the Six Nations but showed genuine concern for all Indians. It would have been consistent for him to have opposed any such scheme as Lyman's, even if it focused, as initially it did not, exclusively on an area inhabited by southern Indians.

23. Joseph Trumbull to William S. Johnson, n.d., but probably 1769. Ibid., 3:57.

24. *New York Journal,* 7 January 1773.

25. Boyd, *Susquehannah Papers,* 4:25.

26. CO5/607:216–19.

27. See Arthur Young, *Political Essays concerning the Present State of the British Empire* (London: Printed for W. Strahan and T. Cadell, 1772); *New York Mercury,* 10 February 1772; *Maryland Gazette,* 3 December 1772; and *Gentleman's Magazine* 42 (1772):63, 355, 509, for representative samples. Young argued that the new lands would produce much-needed hemp, flax, and tobacco. The *Mercury* quoted a report from London of 16 November 1771: "The affair of the Mississippi government is revived since Lord Hillsborough's return from Ireland, and we are told that a civil establishment on that river will take place in the course of the present winter." The *Gazette* too was optimistic: "So great is the passion for lands on the Mississippi in North America, that one petition only, now lying before the Privy Council, has 180 names to it, many of them persons of large fortune, who are already sending to the neighborhood of the Rhine and other parts of Germany for emigrants."

28. Davies, *Documents of the American Revolution,* 6:56.

29. *Pennsylvania Gazette,* 21 October 1772; *Providence Gazette,* 17 October 1772.

30. Historical Manuscripts Commission, *Various Collections* (Dublin: H.M.S.O., 1909), 6:253–54.

31. *Providence Gazette,* 3 October 1772.

32. *Pennsylvania Gazette,* 7 October 1772.

33. *Dictionary of National Biography,* s.v. "Wills Hill, Earl of Hillsborough."

34. Alvord and Carter, *Trade,* p. 241.

35. Chester to Hillsborough, 29 September 1770, 23 June 1771; James Wright to the Board of Trade, 27 December 1771, Davies, *Documents of the American Revolution,* 2:193; 3:123, 278.

36. Gage to Barrington, 5 August 1772, cited in Gipson, *British Empire,* 11:480.

37. Lyman's petition for royal bounty of 10 April 1772 was accompanied by testimonials from Lords Loudoun and Albemarle and from Generals Amherst, Monckton, and Webb. Davies, *Documents of the American Revolution,* 4:67.

38. Gipson, *British Empire,* 11:465, 473.

39. Ibid., p. 441.

40. Barbara Solomon, ed., Timothy Dwight, *Travels in New England and New York* (New York, 1823; reprint, Cambridge, Mass.: Belknap Press of the Harvard University Press, 1969), 1:225.

41. *Pennsylvania Gazette,* 9 December 1772.

42. *Providence Gazette,* 10 October 1772.

43. Benjamin Franklin to William Franklin, 13 June 1767, Alvord and Carter, *The New Regime,* p. 574.

44. *New York Gazette,* 16 November 1772.
45. Ibid., 28 December 1772.
46. *New York Journal,* 7 January 1773.
47. *Providence Gazette,* 30 January 1773.
48. Bates, *Two Putnams,* p. 144.
49. *New York Gazette,* 18 January 1773.
50. Ibid., 25 March 1773.
51. *Providence Gazette,* 15, 22 May, 12 June 1773.
52. Bates, *Two Putnams,* pp. 130, 149, 155.
53. *New York Gazette,* 7 June 1773.
54. Chester to Hillsborough, 23 June 1771, Davies, *Documents of the American Revolution,* 3:123.
55. *New York Gazette,* 24 May 1773.
56. Bates, *Two Putnams,* p. 129.
57. CO5/634, Council Minutes for 22 October 1772.
58. CO5/630, Council Minutes for 5 March 1773.
59. Bates, *Two Putnams,* p. 130.
60. *New York Gazette,* 31 May 1773.
61. Chester to Dartmouth, 16 May 1773, Davies, *Documents of the American Revolution,* 6:145–46.
62. *Providence Gazette,* 24 April 1773.
63. Bates, *Two Putnams,* p. 158.
64. Ibid., p. 131.
65. Ibid., p. 145.
66. Ibid., p. 167.
67. Ibid., p. 166.
68. The treaty is quoted in full in Romans, *Concise History,* pp. 340–42.
69. Almeda Ruth King, "Social and Economic Life in Spanish Louisiana, 1763–1783" (Ph.D. diss., University of Illinois, 1931).
70. Bates, *Two Putnams,* p. 167.
71. Landry, *Voyage to Louisiana,* pp. 101–2.
72. *Providence Gazette,* 10 July 1773.
73. Bates, *Two Putnams,* p. 170.
74. *MPAED,* 8:289–90.
75. Johnson, *British West Florida,* p. 140.
76. Chester to Hillsborough, 29 September 1770, Davies, *Documents of the American Revolution,* 2:189–90.
77. Library of Congress, West Florida Papers, Council Minutes for 14 October 1771.
78. Sosin, *Revolutionary Frontier,* p. 62.
79. CO5/630, Council Minutes for 19 April 1773.
80. *Rivington's New York Gazette,* 7 October 1773.
81. *Providence Gazette,* 24 July 1773.

82. CO5/591:183.

83. Bates, *Two Putnams*, pp. 192, 202–3.

84. Romans, *Concise History*, p. 339.

85. Bates, *Two Putnams*, p. 220.

86. Ibid., pp. 132, 187.

87. *Maryland Gazette*, 19 August 1773.

88. CO5/634:117–18.

89. Ibid., pp. 121–24.

90. Bates, *Two Putnams*, p. 232.

91. *New York Gazette*, 28 June 1773.

92. Ibid., 24 February 1774.

93. *Maryland Gazette*, 19 August 1773.

94. *New York Gazette*, 31 January 1774.

95. Matthew Phelps, *Memoirs and Adventures of Captain Matthew Phelps, Particularly in Two Voyages from Connecticut to the River Mississippi, 1773–1780* (Bennington, Vt.: Haswell, 1802), p. 16 and appendix, p. 60.

96. Bates, *Two Putnams*, p. 45.

97. *New York Journal*, 13 January 1774.

98. Phelps, *Memoirs*, p. 60.

99. CO5/613:290.

100. *Providence Gazette*, 23 July 1774.

101. CO5/613:208–10.

102. CO5/630, Council Minutes for 5 March 1774.

103. "The Answer of Governor Chester to the Petition of Complaint of Adam Christie and Others," in Chester to Germain, 18 February 1781, Library of Congress, West Florida Papers.

104. CO5/630, Council Minutes, 5 March 1774.

105. Barbour Abstracts from Connecticut Town Records, Connecticut State Archives, Hartford, Conn.

106. Franklin Bowditch Dexter, *Biographical Sketches of the Graduates of Yale College with Annals of the College History, October 1701 to September 1815* (New York: Holt, 1885–1912), 3:35.

107. Solomon, *Travels in New England*, 1:225.

108. CO5/607:219.

109. Dwight, *Descendants*, 1:124.

110. Solomon, *Travels in New England*, 1:225.

111. National Archives, General Land Office (Division D).

112. CO5/607:218–26.

113. CO5/617:280.

114. CO5/635, Council Minutes for 27 November 1778.

115. CO5/580:327.

116. Ruth C. Connor, "Gentleman Phil: Eighteenth-Century Opportunist, Philip Peter Livingston, 1740–1810" (M.A. thesis, Auburn University, 1982).

117. Library of Congress, West Florida Papers.

118. National Archives, General Land Office (Division D).

119. Dwight, *Descendants*, 1:131–34; *Dictionary of American Biography*, s.v. "Dwight, Timothy."

120. Phelps, *Memoirs*, pp. 69–71.

121. CO5/635, Council Minutes for 13 November 1778.

122. CO5/631, Council Minutes for 1 September 1777.

123. Claiborne, *Mississippi*, p. 109.

124. National Archives, General Land Office (Division D).

125. Phelps, *Memoirs*, p. 144.

126. CO5/631, Council Minutes for 1 September 1777.

127. Phelps, *Memoirs*, p. 144.

128. CO5/635, Council Minutes for 13 November 1778.

129. Phelps, *Memoirs*, pp. 102–3.

130. CO5/631, Council Minutes for 8 July 1776.

131. The story of the Willing raid is well told in chapter 3 of Haynes, *Natchez District*.

132. *Rivington's New York Gazetteer*, 13 July 1775.

133. "The Answer of Governor Chester."

134. Chester to Germain, 7 May 1778, CO5/594:504.

135. Claiborne, *Mississippi*, pp. 127–29.

136. Those who surrendered to the Spanish included Blommart, Alston, Lieutenant John Smith, Captain Jacob Winfree, William Eason, Parker Carradine, and George Rapalje. Initially taken to New Orleans for trial, they were subsequently pardoned. Peter A. Brannon, "The Coosa Crossing of British Refugees, 1781," *Alabama Historical Review* 19 (1957): 152. One of the two who survived flight into the Choctaw country was Colonel Anthony Hutchins, who reached Savannah and ultimately London.

137. Solomon, *Travels in New England*, 1:91, 226–27.

138. Dwight, *Descendants*, 1:214–15.

139. Solomon, *Travels in New England*, 1:226–29.

140. Dwight, *Descendants*, 1:218–19.

141. Dexter, *Graduates of Yale*, 1:274.

142. Solomon, *Travels in New England*, 1:229.

143. Dwight, *Descendants*, 1:125.

144. National Archives, General Land Office (Division D).

145. Robert J. Taylor, *Colonial Connecticut* (Millwood, N.Y.: KTO Press, 1979), p. 233. In an earlier era the river had been the dividing line between Old Lights and New Lights. Richard L. Bushman, *From Puritan to Yankee* (Cambridge, Mass.: Harvard University Press, 1969), p. 258.

146. Phelps, *Memoirs*, pp. 60–62.

147. Taylor, *Colonial Connecticut*, p. 231.

148. Livermore, *Land Companies*, p. 38.

149. Phelps, *Memoirs*, p. 61.

150. Taylor, *Colonial Connecticut*, p. 227.

151. Boyd, *Susquehannah Papers,* 1:80.

152. Dwight, *Descendants,* 1:218–19.

153. Clifford K. Shipton, *Biographical Sketches of Those Who Attended Harvard in the Classes 1768–1771* (Boston: Massachusetts Historical Society, 1975), 17:144.

154. Connecticut Archives, Revolutionary War, Series 1, 13:144.

155. Guy Carleton to Lord North, 29 November 1783, CO5/111:229.

156. The document is reproduced in Kennet Scott, ed., *Journal of Mississippi History* 26 (1964): 45–46.

157. See my "Bernard Lintot."

158. Oscar Zeichner, *Connecticut's Years of Controversy, 1750–1776* (Williamsburg, Va., 1949; reprint, Hamden, Conn.: Archon Books, 1970), pp. 143–45.

159. Walter Lowrie and Matthew St. Clair Clarke, eds., *American State Papers: Public Lands* (Washington, D.C.: Gales and Seton, 1832), 1:173.

8: Conclusion

1. Chester to Germain, 24 November 1780, CO5/596:73.

2. Browne to Hillsborough, 1 July 1768, CO5/585:115.

3. Same to the same, 25 August 1768, ibid., p. 274.

4. CO5/511:115.

5. PRO T64/276.

6. Alexander Cluny, *The American Traveller; or, Observations on the Present State, Culture, and Commerce of the British Colonies in America* (Philadelphia: Crukshank and Collins, 1770), p. 80, and John Entick, *The Present State of the British Empire* (London: Printed for Law, 1775), 4:474.

7. T64/276.

8. David MacPherson, *Annals of Commerce, Manufactures, Fisheries, and Navigation. . .* (London: Printed for Nichols and Son, 1805), 3:410–23; 4:40.

9. Browne to Hillsborough, 25 August 1768, CO5/585:274.

10. CO5/580:13.

11. Entick, *Present State,* 4:479.

12. "A British Merchant," in *South Carolina Gazette,* 11 April 1768.

13. Gary M. Walton and James F. Shepherd, *The Economic Rise of Early America* (Cambridge: Cambridge University Press, 1979), p. 101.

14. T1/463, pt. 4, fol. 232.

15. T1/479:201.

16. T1/485:226, 228.

17. Haldimand to David Waugh, 24 February 1768, T1/463, pt. 1, fol. 5.

18. T1/475:232.

19. T1/463:118.

20. CO5/68:257.

21. Din, *Louisiana in 1776*, p. 62.

22. Jacob Blackwell to Customs Commissioners, 25 March 1775, T1/505:308.

23. T1/479, pt. 3, fol. 174.

24. Stuart to Botetourt, 12 July 1770, Davies, *Documents of the American Revolution*, 2:147.

25. John Thomas to Williamson, 16 June 1765, Historical Manuscripts Commission, *Report on the Rutland MSS* (London: H.M.S.O., 1888–1894), 4:333; Romans, *Concise History*, pp. 113–14.

26. Van Doren, *Travels of Bartram*, p. 332; Durnford to Hillsborough, 12 May 1769, CO5/587:145, Rowland, *Dunbar*, pp. 28–38.

27. *South Carolina Gazette*, 1 October 1764 through 12 April 1770.

28. Bodleian MS North a. 2, fols. 7, 15–19.

29. *Scots Magazine*, 1764; John Q. Anderson, "The Narrative of John Hutchins," *Journal of Mississippi History* 20 (1958):4.

30. Anderson, "Narrative of John Hutchins," p. 5; Romans, *Concise History*, p. 112; Claiborne, *Mississippi*, p. 115.

31. CO5/580:113.

32. Robert Wells, *Register, Together with an Almanack for the Year of Our Lord, 1775* (Charleston: Wells, 1775).

33. Charles L. Mowat, *East Florida as a British Province, 1763–1784* (Berkeley: University of California Press, 1943), p. 156.

34. *New York Gazette*, 17 March 1766; *New York Journal*, 4 December 1766; *South Carolina Gazette*, 9 February, 15 June 1767.

35. Romans, *Concise History*, p. 192.

36. Ibid., p. 106, and Phineas Lyman's "Reasons," in Alvord and Carter, *The New Regime*, p. 307.

37. Romans, *Concise History*, p. 83.

38. Daniel Hickey to Montfort Browne, 7 April 1771, T1/494:57.

39. Romans, *Concise History*, p. 111.

40. Library of Congress, West Florida Papers, Council Minutes for 8 June 1771.

41. Romans, *Concise History*, pp. 104, 204.

42. U.S. Dept. of Commerce, *Historical Statistics*, 2:1176.

43. Walton and Shepherd, *Economic Rise*, p. 196.

44. Gray, *History of Agriculture*, 1:137, 2:1024.

45. U.S. Dept. of Commerce, *Historical Statistics*, 2:1176.

46. Johnson, *British West Florida*, p. 96.

47. CO5/577:76.

48. Howard, *British Development*, p. 87.

49. Rea and Howard, *General Assembly*, p. xxii.

50. CO5/612:270–78.

51. Rea, "Planters," p. 231n.

52. Arthur, *Despatches of the Governors*, 2:193.

53. Library of Congress, West Florida Papers, Council Minutes for 1 May 1770.

54. Rea and Howard, *General Assembly*, pp. xxiv, xxv.

55. Dalrymple, *Merchant of Manchac*, pp. 425–32.
56. CO5/599:173.
57. Chester to Hillsborough, 27 January 1771, CO5/587:87.
58. CO5/577:80.
59. General John Campbell to Germain, 10 May 1779, *MPAED*, 9:84–85.
60. CO5/591:183.
61. CO5/577:71.
62. CO5/613:21–26.
63. National Archives, General Land Office (Division D).

Bibliographic Essay

For anyone researching the economic or indeed any aspect of British West Florida, the prime source of information is the set of sixty-two manuscript volumes in the Colonial Office's West Florida papers in Great Britain's Public Record Office at Kew (CO5/574–635), of which there are copies in the Library of Congress in Washington. Here the set on microfilm in the Ralph Brown Draughon Library at Auburn was used. Volumes of particular relevance were CO5/574–75, 577–93 (Governors' Correspondence); 576, 601–4, 606–12 (Land Petitions, Grants, and Records); 613–17 (Registered Legal Documents); 595–96, 599, 600, 619 (Letters from the London Government); and 630–35 (Council Minutes, 1764–1780). Supplementing these sources were ten manuscript volumes on three reels of microfilm from the U.S. National Archives in Washington made in 1968 for the director of libraries at the University of West Florida, Pensacola, which include overlapping and extra information about landholding in West Florida. A special order and not included in the National Archives catalog, they are from the Department of the Interior, Bureau of Land Management, Records of the General Land Office, Record Group 49, Division D (Private Land Claims).

A quantity of the Public Record Office material as well as pertinent selections from the Cuban papers in the Archive of the Indies in Seville, Spain, were transcribed earlier in this century by Dunbar Rowland and are housed in the Mississippi State Archives in Jackson. The original intention to publish them as *Mississippi Provincial Archives: English Dominion* and *Mississippi Provincial Archives: Spanish Dominion* in ten and eight volumes respectively has so far resulted in only the first volume of the Public Record Office transcripts, which was edited by Rowland and printed in 1911 in

Nashville by the Brandon Printing Company. Also in the Public Record Office itself is the vast collection of Treasury Board 1 (In-Letters), which has yet to be fully explored for West Florida materials. I have found pertinent information in volumes 463, 471, 475, 479, 485, 494, and 505. Treasury Board 64 (Miscellaneous) was also useful. At Kew again are the accounts of the Audit Office. Of particular interest are numbers 126 and 127, which refer to the accounts of Governor Johnstone and John Ellis, the colonial agent for West Florida. Customs transactions concerning West Florida are unknown and probably cannot be known in detail, but Cust. 16/1, 17/2, 17/3, 17/4, 17/5, and 17/6, housed at the Public Record Office, refer to the imports and exports of both the Floridas. In this book most material derived from them came from transcripts kindly loaned by Robert Rea. Additional and more detailed information on West Floridian imports and exports, although for fewer years, can be found in the North MSS in the Bodleian Library, Oxford, England. Of value in plumbing the considerable impact of the military on the economic life of West Florida are the British Headquarters Papers kept at the library of Colonial Williamsburg, Inc., Williamsburg, Va., of which Auburn University possesses microfilm copies. Useful too are some of the Shelburne Papers in the William Clements Library, Ann Arbor, Mich. For this book photostats of those papers relating to West Florida in the library of Florida State University at Tallahassee were used.

For the activities of Montfort Browne, lieutenant governor of West Florida, the papers of the earl of Dartmouth in the Staffordshire County Records Office, Stafford, were available. For biographical details of the immigrants to West Florida who belonged to the Company of Military Adventurers, the Connecticut Archives Manuscript Division, Revolutionary War, 1763–1789, Series 1, proved useful. I am grateful to Auburn University for a grant-in-aid that allowed me to see them. For a transcript of the Reverend William Gordon's report on West Florida's economic development in the Ellis Papers in the Linnaeus Library, London, I am grateful to Roy A. Rauschenberg of Ohio University, and for a transcription of the letter book of Charles Strachan, a merchant of Mobile, which is to be found in the National Library of Scotland, Edinburgh, I am again indebted to Robert Rea.

Besides Strachan's letter book, the only ones surviving of the many merchants who did business in British West Florida are those of John Fitzpatrick. Long available only in manuscript in the New York Public Library, they were edited by Margaret F. Dalrymple and published in 1978 in Baton Rouge by the Louisiana State University Press as *The Merchant of Manchac: The Letterbooks of John Fitzpatrick, 1768–1790*. They are invaluable as a guide to the commercial connections among West Floridians and to the role of New

Orleans in the economy of the province. Among document collections of great relevance and convenience is the printed compilation of *The Minutes, Journals, and Acts of the General Assembly of British West Florida*, the work of Robert R. Rea and the late Milo B. Howard, Jr., which was published by the University of Alabama Press at University, Ala., in 1979. Wider in scope but more limited in the years covered is Kenneth G. Davies, *Documents of the American Revolution, 1770–1783*, 21 vols. (Shannon: Irish University Press, 1972–1981), which reproduces in extenso many documents relating to West Florida from British government sources and lists even more.

Sidelights on the maritime life of West Florida during the revolution, using American sources too, may be found in *Naval Documents of the American Revolution* (Washington, D.C.: U.S. Government Printing Office, 1964–), of which eight volumes have so far been printed. The first four volumes were edited by William Bell Clark and the remainder by William James Morgan. Documents relating to the trade of West Florida are to be found in the *Collections of the Illinois State Historical Society*, most particularly in three volumes edited by Clarence W. Alvord and Clarence E. Carter: *The Critical Period, 1763–1765; The New Regime, 1765–1767;* and *Trade and Politics, 1767–1769*. They were all published in Springfield, Ill., by the Illinois State Historical Society, in 1915, 1916, and 1921 respectively. Some of the material in them is also found in Carter's *The Correspondence of General Thomas Gage with the Secretaries of State and with the War Office and the Treasury, 1763–1775*, 2 vols. (New Haven, 1933; reprint, New York: Archon Books, 1969). Other documents relating to West Florida may be examined in volumes 1 and 5 of *Publications of the Mississippi Historical Society*, Centenary Series, 5 vols. (Jackson, Miss.: Press of the Mississippi Historical Society, 1916–1925).

Other document collections informative of the economic life of West Florida are volumes 11–14 of *Journal of the Commissioners for Trade and Plantations* (London: His Majesty's Stationery Office, 1935–1938) and W. L. Grant and James Munro, eds., *Acts of the Privy Council*, Colonial Series, vols. 4–6 (London: Lords Commissioners of the Treasury, 1908–1912). Several calendars from the British Historical Manuscripts Commission are still handy. *The Report on American Manuscripts in the Royal Institution of Great Britain*, 4 vols. (London: H.M.S.O., 1904–1909), is a guide to the previously mentioned British Headquarters Papers. The commission's *Fifth Report*, pt. 1 (London: H.M.S.O., 1876), *Sixth Report* (London: H.M.S.O., 1877), and *Report on the Rutland MSS*, 4 vols. (London: H.M.S.O., 1888–1894), also have their uses. On the economic relations between West Florida and her Spanish neighbor, the *Despatches of the Spanish Governors of Louisiana*, published in New Orleans in 8 volumes between 1937 and 1938 for the Works Progress Administration, edited by Stanley C. Arthur and translated into English by Carmen Philpott and Marion Scineaux, illumines an unfamiliar subject. Volume 1 of Arthur Berriedale Keith, ed., *Selected Speeches and Documents on British Colonial Policy, 1763–1917*, 2 vols. (London: Oxford University Press, 1933), is also valuable as a convenient place to consult the important proclamation of 1763.

Limited printed primary sources also include *The Log of H.M.S. Mentor*, edited by James A. Servies and with an introduction by Robert R. Rea (Pensacola: University Presses of Florida, 1982), and the diary and letters of a Baton Rouge planter, in Eron D. Rowland's *Life, Letters, and Papers of William Dunbar* (Jackson, Miss.: Press of the Mississippi Historical Society, 1930), as well as *William Bartram's Travels through North and South Carolina, Georgia, East and West Florida*, edited by Mark Van Doren and republished in 1955 in New York by Dover Publications from the 1791 Philadelphia edition. Bartram was a naturalist, and Bernard Romans had similar interests, but his *Concise Natural History of East and West Florida* (New York, 1775; reprint, Gainesville: University of Florida Press, 1962) is concerned with much else besides and is excellent in detailing the difficulties and expenses of immigration to West Florida. A rather earlier visitor to West Florida was Lord Adam Gordon, whose "Journal of an Officer Who Travelled in America and the West Indies in 1764 and 1765" may be found in Newton D. Mereness, ed., *Travels in the American Colonies* (New York, 1916; reprint, New York: Antiquarian Press, 1961), pp. 367–453. Thomas Hutchins, an actual resident of West Florida, published *An Historical Narrative and Topographical Description of Louisiana and West Florida* in Philadelphia in 1784, but it is now more easily available in a facsimile edition published in Gainesville in 1968 by the University of Florida Press. Contemporary descriptions of the western region of West Florida can be found in Philip Pittman, *The Present State of the European Settlements in the Mississippi*, intro. Robert Rea (London, 1770; reprint, Gainesville: University of Florida Press, 1973), and in Albert C. Bates, ed., *The Two Putnams: Israel and Rufus in the Havana Expedition, 1762, and in the Mississippi Exploration, 1772–1773, with Some Account of the Company of Military Adventurers* (Hartford: Connecticut Historical Society, 1931). One of the Military Adventurers was Matthew Phelps, whose *Memoirs and Adventures of Captain Matthew Phelps, Particularly in Two Voyages from Connecticut to the River Mississippi, 1773–1780* (Bennington, Vt.: Haswell, 1802) badly needs to be edited and reprinted.

More about the Military Adventurers may be read in Barbara Solomon, ed., Timothy Dwight, *Travels in New England and New York*, 4 vols. (New York, 1823; reprint, Cambridge, Mass.: Belknap Press of Harvard University Press, 1969). There is information on how their descendants vainly tried to recover land they had settled in Walter Lowrie and Matthew St. Clair Clarke, eds., *American State Papers: Public Lands,* vol. 1 (Washington, D.C.: Gales & Seton, 1832). Some insights into the Indian trade may be obtained from James Adair's rambling and poorly titled *History of the American Indians* (London, 1775; reprint, Johnson City, Tenn.: National Society of the Colonial Dames of America, 1930). Peripheral information about West Florida may be gleaned from Charles César Robin's *Voyages dans l'intérieur de la Louisiane, de la Floride occidentale et dans les Isles de la Martinique et de Saint Domingue pendant les années 1802, 1803, 1804, 1805, et 1806* (Paris: Buisson, 1807), volumes 2 and 3 of which were translated and edited by Stuart A. Landry. They were published in abridged form as *Voyage to Louisiana* in New Orleans in 1966 by the Pelican Printing Company. More pertinent is Gilbert C. Din., ed., *Louisiana in 1776: A Memoir of Francisco Bouligny* (New Orleans: Louisiana Collection Series, 1977), which has much on the trade connection with West Florida. M. Le Page du Pratz, *The History of Louisiana* (London, 1774; reprint, Baton Rouge: Louisiana State University Press, 1975), is useful for details of contemporary farming techniques. Odd items of interest about West Florida may also be found in Thomas Whateley, *The Regulations Lately Made concerning the Colonies and the Taxes Imposed upon Them* (London: Printed for Wilkie, 1765); Arthur Young, *Political Essays concerning the Present State of the British Empire* (London: Printed for Strahan and Cadell, 1772); volume 7 of George C. Rogers, Jr., David R. Chesnutt, and Peggy J. Clark, eds., *The Papers of Henry Laurens* (Columbia, S.C.: University of South Carolina Press, 1977); John C. Fitzpatrick, ed., *The Writings of George Washington,* 39 vols. (Washington, D.C.: U.S. Government Printing Office, 1931–1944); and from the microprint publication series *Early American Imprints, 1639–1800,* carried out under Clifford K. Shipton's direction at the American Antiquarian Society, Worcester, Mass., especially number 10252.

Contemporary if rather suspect statistics on the volume and value of British West Florida's trade may be read in Alexander Cluny, *The American Traveller; or, Observations on the Present State, Culture, and Commerce of the British Colonies in America* (Philadelphia: Crukshank and Collins, 1770); John Entick, *The Present State of the British Empire,* 4 vols. (London: Printed for Law, 1775); and in volumes 3 and 4 of David MacPherson, *Annals of Commerce, Manufactures, Fisheries, and Navigation . . .* 4 vols. (London: Printed for Nichols and Son, 1805). Robert Wells's *Wells' Register Together with an Almanack for the Year of Our Lord, 1775* (Charleston: Wells, 1775) is chiefly useful as a guide to the personnel of the civil and military establishment of West Florida in the mid-1770s, but it also gives, most usefully, the exchange rates of sterling with the Spanish and Portuguese coinage then current in West Florida. Though they must be used cautiously, since rumors are sometimes reported as facts, newspapers have been unjustifiably neglected by students of West Florida. Very often their news was about economic matters. Those most useful are the *Boston Chronicle, Caledonian Mercury, Jamaica Mercury* (of which only short and incomplete runs have survived in the British Library Newspaper Collection, Colindale, London), *London Gazette, London Magazine, Maryland Gazette, Massachusetts Gazette and Boston Newsletter, New York Journal, New York Mercury, Pennsylvania Gazette, South Carolina Gazette, South Carolina and American General Gazette,* and the *Scots Magazine.*

An early secondary source which can still be useful is Peter J. Hamilton, *Colonial Mobile* (Boston, 1910; reprint, University, Ala.: University of Alabama Press, 1976), but secondary works of fundamental importance include two from the 1940s: Cecil Johnson, *British West Florida, 1763–1783* (New Haven: Yale University Press, 1942), and Clinton N. Howard, *The British Development of West Florida, 1763–1769* (Berkeley: University of California Press, 1947). Of vital use for perspective and comparison with its sister province is Charles L. Mowat, *East Florida as a British Province, 1763–1784* (Berkeley: University of California Press, 1943). Volume 9 in Lawrence H. Gipson's magisterial 15-volume work *The British Empire before the American Revolution* (New York: Knopf, 1958–1970) is also excellent on the early years of British West Florida's existence. Only volume 1 of J. F. H. Claiborne's intended 2-volume work *Mississippi as a Province, Territory, and State* (Jackson, Miss., 1880; reprint, Baton Rouge: Louisiana State University Press, 1964) exists, but it covers the British period and contains material not found elsewhere. Lewis C. Gray, *History of Agriculture in the Southern United States,* 2 vols. (New York: Smith, 1941), contains a great deal of esoteric information on farming methods in the eighteenth century, although Gray unfortunately does not deal with East and West Florida separately.

Also useful for economic aspects of the province are United States Department of Commerce, Bureau of the Census, *Historical Statistics of the United States: Colonial Times to 1970,* 2 vols., with an introduction by Ben J. Wattenberg (New York: U.S. Government Printing Office, 1976), and Gary M. Walton and James F. Shepherd, *The Economic Rise of Early America* (Cambridge: Cambridge University Press, 1979). On Britain's trade policies toward the American colonies, Oliver M. Dickerson, *The Navigation Acts and the American Revolution* (Philadelphia, 1951; reprint, New York: Barnes, 1963); Carl Ubbelohde, *The Vice Admiralty Courts and the American Revolution* (Chapel Hill: University of North Carolina Press, 1960); Ian R. Christie and Benjamin W. Labaree, *Empire or Independence, 1760–1776* (New York: Norton, 1976); and Frances Armytage, *The Free Port System in the British West Indies*

(London: Longmans, Green, 1953), all provide background for the case of West Florida. On its maritime life William Smith, *History of the Post Office in British North America* (Cambridge, 1920; reprint, New York: Octagon Books, 1973); Richard B. Sheridan, *Sugar and Slavery: An Economic History of the British West Indies, 1623–1775* (Baltimore: Johns Hopkins University Press, 1973); and John D. Ware and Robert R. Rea, *George Gauld, Surveyor and Cartographer of the Gulf Coast* (Gainesville: University Presses of Florida, 1982), likewise all have relevance. For the mysteries of eighteenth-century nautical language and lore, both William Falconer, *An Universal Dictionary of the Marine* (London, 1780; reprint, New York: Kelley, 1970), and Peter Kemp, ed., *The Oxford Companion to Ships and the Sea* (London: Lane, 1976), are helpful.

For West Florida during the years of revolution and war, the best secondary books are J. Barton Starr, *Tories, Dons, and Rebels* (Gainesville: University Presses of Florida, 1976); Robert V. Haynes, *The Natchez District in the American Revolution* (Jackson, Miss.: University Press of Mississippi, 1976); J. W. Caughey, *Bernardo de Gálvez in Louisiana, 1776–1783* (Berkeley: University of California Press, 1934); and F. de Borja Medina Rojas, *José de Ezpeleta, Gobernador de la Mobila, 1780–1781* (Seville: Escuela de Estudios Hispano-Americanos de Sevilla, 1980). On Indians the key works are John R. Alden, *John Stuart and the Southern Colonial Frontier* (Ann Arbor: University of Michigan Press, 1944); David H. Corkran, *The Creek Frontier, 1540–1783* (Norman, Okla.: University of Oklahoma Press, 1967); Helen L. Shaw, *British Administration of Southern Indians, 1756–1783* (Lancaster, Pa.: Lancaster Press, 1931); and J. Leitch Wright, Jr., *The Only Land They Knew* (New York: Free Press, 1981). Germane to the topic is an old classic, Clarence W. Alvord, *The Mississippi Valley in British Politics,* 2 vols. (New York: Russell & Russell, 1916). Secondary works mentioning West Florida's trade links with New Orleans include John G. Clark, *New Orleans, 1718–1812: An Economic History* (Baton Rouge: Louisiana State University Press, 1970), and Bertram Korn, *The Early Jews of New Orleans* (Waltham, Mass.: American Jewish Historical Society, 1969).

There are brief accounts of the Company of Military Adventurers in Jack Sosin, *The Revolutionary Frontier, 1763–1783* (New York: Holt, Rinehart & Winston, 1967), in Shaw Livermore, *Early American Land Companies* (New York: Commonwealth Fund, 1939), and, rather more fully, in Bernard Bailyn, *Voyagers to the West* (New York: Knopf, 1986). A biography of its leader by Delphina H. Clark entitled *Phineas Lyman, Connecticut's General* was published in Springfield, Mass., in 1964 by the Connecticut Valley Historical Museum. It unfortunately treats his West Floridian experience in a cursory manner.

Other important members of the company receive attention in Benjamin W. Dwight, *The History of the Descendants of John Dwight of Dedham*, 2 vols. (New York: Printed by Trow and Son, 1874), and information on other Adventurers may be gleaned from Julian Boyd, ed., *The Susquehannah Company Papers*, 11 vols. (Ithaca, N.Y.: Cornell University Press, 1930–1971); David Syrett, ed., *The Siege and Capture of Havana, 1762* (London: Navy Records Society, 1970); Franklin Bowditch Dexter, *Biographical Sketches of the Graduates of Yale College with Annals of the College History, October 1701 to September 1815*, 6 vols. (New York: Holt, 1885–1912); and Clifford K. Shipton, *Biographical Sketches of Those Who Attended Harvard in the Classes 1768–1771* (Boston: Massachusetts Historical Society, 1975). The background to the migration from Connecticut can be found in Robert J. Taylor, *Colonial Connecticut* (Millwood, N.Y.: KTO Press, 1979); Richard L. Bushman, *From Puritan to Yankee* (Cambridge, Mass.: Harvard University Press, 1969); and Oscar Zeichner, *Connecticut's Years of Controversy, 1750–1776* (Williamsburg, 1949; reprint, Hamden, Conn.: Archon Books, 1970).

A number of theses and dissertations as well as innumerable articles relate to the subject of this work. Many of them present material covered elsewhere, and only some of the most valuable will be mentioned here. Almeda Ruth King's University of Illinois doctoral dissertation "Social and Economic Life in Spanish Louisiana, 1763–1783," which was finished in 1931, gives valuable information on a neighboring colony during the years of British West Florida's existence. Also of worth is Manuel I. Pérez-Alonso, "War Mission in the Caribbean: The Diary of San Francisco de Saavedra (1780–83)," a Ph.D. dissertation completed in 1953 at Georgetown University. More exclusively concerned with West Florida are four M.A. theses written at Auburn University: Elizabeth May Jones Conover, "British West Florida's Mississippi Frontier during the American Revolution" (1972); Kathryn E. Holland, "The Path between the Wars: Creek Relations with the British Colonies, 1763–1774" (1980); Ruth C. Connor, "Gentleman Phil: Eighteenth-Century Opportunist, Philip Peter Livingston, 1740–1810" (1982), which emphasizes Livingston's eight years as West Florida's provincial secretary and chief land speculator; and Debra L. Fletcher, "They Lived, They Fought: The Creek-Choctaw War, 1763–1776" (1983).

Articles on Indians relevant to their trade, which, incidentally, is intertwined in most of their dealings with white men, include Peter A. Brannon, "The Pensacola Indian Trade," *Florida Historical Quarterly* 31 (July 1952); Clarence E. Carter, "British Policy toward the American Indians in the South, 1763–68," *English Historical Review* 33 (1918); Robin F. A. Fabel and Robert R. Rea, "Lieutenant Thomas Campbell's Sojourn among the Creeks, November

1764–May 1765," *Alabama Historical Quarterly* 36 (Summer 1974); James H. O'Donnell III, "Alexander McGillivray: Training for Leadership, 1777–1783," *Georgia Historical Quarterly* 49 (June 1965); George C. Osborn, "Relations with the Indians in West Florida during the Administration of Governor Peter Chester, 1770–1781," *Florida Historical Quarterly* 31 (April 1953); Robert R. Rea, "Redcoats and Redskins on the Lower Mississippi, 1763–1776: The Career of Lieutenant John Thomas," *Louisiana History* 11 (Winter 1970); and Helen H. Tanner, "Pipesmoke and Muskets: Florida Indian Intrigues of the Revolutionary Era," in Samuel Proctor, ed., *Eighteenth Century Florida and its Borderlands* (Gainesville, 1975), the first in a series of five symposia proceedings published by the Florida Bicentennial Commission to celebrate the revolution. Of a similar genre is William S. Coker and Robert R. Rea, eds., *Anglo-Spanish Confrontation on the Gulf Coast* (Pensacola, 1982), the proceedings of the ninth Gulf Coast History and Humanities Conference, which contains an excellent article on Indians, "The Creek Confederacy in the American Revolution: Cautious Participants," by Michael D. Green, which has a strong economic content.

The same volume contains my "Anglo-Spanish Commerce in New Orleans during the American Revolutionary Era." John W. Caughey's "Bernardo de Gálvez and the English Smugglers on the Mississippi, 1777," *Hispanic-American Historical Review* 12 (1932), deals with an aspect of the same topic. Other relevant material may be found in English translations of the record of court cases in New Orleans published serially in the *Louisiana Historical Quarterly* in the 1920s and 1930s, particularly the volumes for 1923, 1925, 1928, 1929, 1930, and 1938, and in my "Letters of 'R': The Lower Mississippi in the Early 1770s," *Louisiana History* 24 (Fall 1983).

Articles on immigration and settlement tend to be preoccupied with the western part of the province. They include John Q. Anderson, "The Narrative of John Hutchins," *Journal of Mississippi History* 20 (1958); W. M. Drake, "A Note on the Jersey Settlers of Adams County," *Journal of Mississippi History* 15 (1953); Robin F. A. Fabel, "Bernard Lintot: A Connecticut Yankee on the Mississippi, 1775–1805," *Florida Historical Quarterly* 60 (July 1981); Cecil Johnson, "Expansion in West Florida, 1770–1779," *Mississippi Valley Historical Review* 20 (1934); Albert Tate, Jr., "The Spanish Census of the Baton Rouge District for 1786," *Louisiana History* 24 (Winter 1983); and Gordon Wells, "British Land Grants—The Wilton Map of 1774," *Journal of Mississippi History* 28 (1966). A more easterly slant may be read in Clinton N. Howard, "Early Settlers in British West Florida," *Florida Historical Quarterly* 24 (July 1945); J. Barton Starr, "Campbell Town: French Huguenots in British West Florida," *Florida Historical Quarterly* 54 (April 1976); and Robin F. A. Fabel, "James

Thompson, Pensacola's First Realtor," *Florida Historical Quarterly* 62 (July 1983).

More general works on the economy of West Florida are Clinton N. Howard, "Some Economic Aspects of British West Florida, 1763–1768," *Journal of Southern History* 6 (May 1940), and Robert R. Rea's articles "Planters and Plantations in British West Florida," *Alabama Review* 29 (July 1976), and "British West Florida Trade and Commerce in the Customs Records," *Alabama Review* 37 (April 1984), the only existing article which deals at all with the provinces' maritime life. Douglas Brown dealt with the biggest single scheme designed to achieve prosperity in West Florida in "The Iberville Canal: Its Relation to Anglo-French Commercial Rivalry, 1763–1775," *Mississippi Valley Historical Review* 32 (1946), while less ambitious projects feature in Robin F. A. Fabel, "George Johnstone and the 'Thoughts concerning Florida'—A Case of Lobbying?" *Alabama Review* 19 (July 1976). The gritty reality of trying to make money in the province can be seen in James D. Born, Jr., "Charles Strachan in Mobile: The Frontier Ordeal of a Scottish Factor, 1764–1768," *Alabama Historical Quarterly* 27 (Spring & Summer 1965).

Views of West Florida from the other side of the Atlantic have not received thorough treatment in articles, but the most celebrated of the province's agents in London has been the subject of two: Robert R. Rea, "The King's Agent for British West Florida," *Alabama Review* 16 (April 1963), and Roy A. Rauschenberg, "John Ellis, Royal Agent for West Florida," *Florida Historical Quarterly* 62 (July 1983). The effect of the most notorious British statute of the 1760s is discussed both in Wilfred B. Kerr, "The Stamp Act in the Floridas, 1765–1766," *Mississippi Valley Historical Review* 21 (1935), and more thoroughly regarding West Florida in J. Barton Starr, "The Spirit of What Is There Called Liberty: The Stamp Act in British West Florida," *Alabama Review* 29 (October 1976). Donna J. Spindel's "The Stamp Act Crisis in the British West Indies," *Journal of American Studies* 11 (August 1977), adds perspective through consideration of other colonies which acquiesced in the act.

For other articles and books on West Florida in the British period, the interested reader should turn to James A. Servies, *Bibliography of West Florida*, 4 vols. (Pensacola: University of West Florida, 1982).

Index

About the Author

Robin F. A. Fabel received his bachelor's and master's degrees from Oxford University and his doctorate from Auburn University. He is an associate professor of history at Auburn, and the author of *Bombast and Broadsides,* a biography of West Florida governor George Johnstone.